THEORY
OF
QUADSITRON-ENERGY
CONNECTIVITY

THEORY
OF
QUADSITRON-ENERGY
CONNECTIVITY

Lane B. Scheiber II, MD
Lane B. Scheiber, ScD

THEORY OF QUADSITRON-ENERGY CONNECTIVITY

iUniverse books may be ordered through booksellers or by contacting:

iUniverse
1663 Liberty Drive
Bloomington, IN 47403
www.iuniverse.com
1-800-Authors (1-800-288-4677)

ISBN: 978-1-5320-7149-2 (sc)
ISBN: 978-1-5320-7150-8 (e)

Library of Congress Control Number: 2019903372

Print information available on the last page.

iUniverse rev. date: 05/21/2019

VIReSOFT collaborating with MedStar Labs, developers of Medically Therapeutic RNA Vector Technologies, Medical Vector Therapy, Molecular Virus Killers, Quantum Gene, Executable Gene, Genetic Reference Tables, Prime Genome, Prime Genomic Cube, Genomic Keycode, Essential Equation 4 Life, Dandelion Rift, the Tritron, the Quadsitron, the Quadsistor, Fourth Generation Biologics and Molecular Gene Activators, Embedded DNA Vaccines, Courier Gene Technology, Theory of the Quadsitron Ether, Theory to Unify GLEAM^2 (Gravity, Light, Electrons, Atoms, Molecules, Magnetism), Doreen Lightspeed Interstellar Gravity Hypercoil Turbine (DORELIGHT) Engines.

MedStar Labs, Inc.

As the author, I would like to state to the reader that the wordsmithing of this publication may have been better, but three very serious health events suddenly threatened my life, which led me to expedite this project. Reality stared me in the eyes. Concerned this text might never reach completion, I moved to publish as soon as reasonably possible. The sole objective of this work is to set fire one's imagination; such as a brilliant sunrise might stimulate the senses, to set in motion the study and discovery of the laws of physics in a new and innovative way. Much work remains to be done. Thank you for your understanding. Best wishes. Lane B. Scheiber II

DNA VACCINES
Courier Gene Technology
Changing the Global Approach to Medicine Series, Volume 5
by Lane B. Scheiber II, MD and Lane B. Scheiber, ScD

FOURTH GENERATION BIOLOGICS: Molecular Virus Killers
Changing the Global Approach to Medicine Series, Volume 4
by Lane B. Scheiber II, MD and Lane B. Scheiber, ScD

CHANGING THE GLOBAL APPROACH TO MEDICINE, Volume 3
Cellular Command and Control
Also introducing the Prime Genome and the Tritron
by Lane B. Scheiber II, MD and Lane B. Scheiber, ScD

CHANGING THE GLOBAL APPROACH TO MEDICINE, Volume 2
Medical Vector Therapy
Also introducing the Quantum Gene and the Quadsistor
by Lane B. Scheiber II, MD and Lane B. Scheiber, ScD

CHANGING THE GLOBAL APPROACH TO MEDICINE, Volume 1
New Perspectives on Treating AIDS, Diabetes, Obesity, Aging, Heart Attacks, Stroke, and Cancer
by Lane B. Scheiber II, MD and Lane B. Scheiber, ScD

IMMORTALITY: QUATERNARY MEDICINE CODE
by Anthony Scheiber

CURSE OF THE SNOW DRAGON
by Anthony Scheiber

THE HUMAN COMPUTER
by Anthony Scheiber

EARTH PRO: The Rings of Sol
by Anthony Scheiber

BLACK DIAMOND
By Anthony Scheiber

DEDICATION

Thanks to our wives, Karin and Mary Jane,
for all of their love and support, without which this
effort could never have been accomplished.

Thanks to Keith, my best friend and confidant,
for being with me and my family, over all these years.

Thanks to Steve, Kurt, and Dave, my brothers,
for all of their encouragement and support.

I would especially like to thank all of the
hardworking teachers whom contributed to my
education, these many years.
A great teacher not only educates,
but challenges you to become better,
so that you can see the world in a different way.

THEORY OF THE QUADSITRON-ENERGY CONNECTIVITY IS AS FOLLOWS:

The universe is comprised solely of
two fundamental nonreducible
components,
that of
quadsitrons and quanta of energy...

...from these 'two' principal entities
all else in the universe is
contrived and connected.

KEY DERIVATIONS

Quadsitron calculated mass of $9.5075301706 * 10^{-38}$ kg.

Quadsitron calculated diameter of $8.557941465 * 10^{-19}$ m.

Energy signature of a single quadsitron is $8.54494198 * 10^{-21}$ joules.

A Quanta of free Energy is equivalent to $1.7 * 10^{-20}$ joules.

All else is simply icing on the cake.

A NECESSARY SHIFT
IN PERSPECTIVE #1

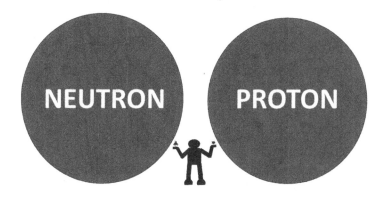

TO FULLY EXPLORE AND appreciate the concepts presented in this text, one must first undergo a frameshift to adjust their perception of the universe we live in. One must think in terms of the world which exists far below the diameter of a neutron or a proton. Inquisitive individuals must take their astute analytical skills and enter the sub-sub atomic realm where quanta of energy and quadsitrons exist. As an analogy, if the smallest entity in the universe, a quadsitron, were a one-centimeter diameter sphere, then a proton would be a sphere with a diameter of 26 meters; in the English measurement system, if a quadsitron were one inch in diameter, a proton would be a mammoth 216 2/3 feet in diameter. It is from this, the smallest of all matter and units of energy, that all of the rest of the universe is constructed. There is a distinct connectivity from the very elemental units of matter to the macro universe human senses have been designed to perceive. The fabric of the space permeates through all of the atoms which comprise our being and surrounds us in a continuum stretching out in all directions, cradling all of the stars and all of the galaxies and nebulas, reaching out to the distant boundaries of the universe, to the very edge of space. The sub-sub atomic fabric, comprised of quadsitrons and energy, is the universe.

NECESSARY SHIFT IN PERSPECTIVE #2

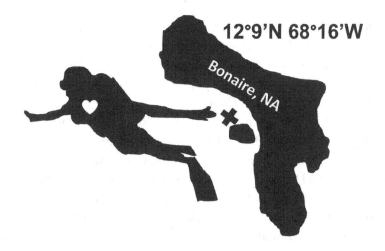

12°9'N 68°16'W

Bonaire, NA

THE STUDY OF SUBatomic physics necessitates re-orienting the human brain to a three-dimensional world. To most, the three-dimensional world which surrounds us is plainly obvious. Standing on a bus corner, or any other part of the planet where the feet are planted firmly on the ground, one may look forward, one may peer to either side, one may gaze behind them or stare up into the sky above. For all practical purposes, the world appears to be three-dimensional habitat. One can convince themselves that they do not need to venture beyond their comfort zone of living in this space in order to be a student of physics.

But the nature of physics is three-dimensional in the purest form. To truly study the operational aspects of physics, one must first place themselves in a three-dimensional medium to allow their brain to appreciate the functions of physics in three-dimensional space. Cracking physics problems utilizing a litany of equations scratched on the surface of a two-dimensional blackboard, provides the means to model physics, but not to experience living physics. To envision three-dimensional physics, one may imagine floating amongst the clouds, whisking around in space, or diving the deepest ocean. To become aware of physics first hand, the most practical approach may be to participate in recreational scuba diving.

To be a recreational scuba diver entails a certain level of training to allow a person to safely venture below the surface of the water with breathing apparatus and an air tank. Recreational diving is generally to a depth of 130 feet, with most dives being between thirty and sixty feet. Scuba diving allows an individual to swim amongst the sea life which inhabit the submerged ecosystem, covering 72% of the earth's surface. Scuba diving is not for everyone, but for those students of physics whom can venture into this three-dimensional world below the surface of the ocean, the mind becomes uniquely appreciative of the three-dimensional aspects of the world around them.

One of the most exciting and educational experiences in three-dimensional physics is if one happens to venture to Bonaire, 12°9'N 68°16'W. Bonaire is comprised of one mountainous main island spanning approximately 111 square miles, and one smaller flat terrain minor island of about 2.3 square miles. This pair of islands in the southern Caribbean, territory of the Netherlands, is located fifty miles off the coast of Venezuela. Bonaire has a number of activities for the adventurer, but a wide variety of scuba diving venues is what this tropical destination is most noted for when one considers visiting the island.

If one is able to take a boat dive trip and if the dive master is willing to drop your party into the water at the northwestern corner of the small island of Klein Bonaire, the most exciting study of physics takes place. Drift diving from the northwest corner of the smaller island westward along the northern coast line of the minor island into the inlet of Bonaire is as close as an average human might get to flying like a bird. The water is crystal clear. The temperature of the water is often 80 degrees. As the saltwater ocean current sweeps across the underwater terrain of the minor island headed towards the main island, divers on a drift dive bob up and down as the submerged landscape passes below their fins. Steering one-self, carried by the ocean current, truly orients the student's mind to the three-dimensional structure of the universe, which is key in appreciating the principles of physics. One might imagine it is like swimming through a gravity ocean.

EDGE OF SURVIVAL

The Age of the Combustion Engine
has mistakenly persisted far too long;
This scalding fiery metal dragon,
belching a wicked pestilence across the planet
is destroying the viability of the world;
Our understanding of Physics
must change,
by necessity;
must evolve to a higher order.
We must re-awaken the concepts of a luminous aether
and the universe's continuous flow of energy;
Intimately incorporating
these fundamentals into our technologies-
If humans
dare hope to survive
to see the twenty-third century.

CONTENTS

PREFACE

THE THEORY OF QUADSITRON-ENERGY Connectivity, is defined by 'all matter exists as a continuum and is intimately connected'. Continuum and connectivity in this instance refer to an uninterrupted connection, succession or union. Everything in the universe is contrived from the same two sub-sub (sub^2) atomic rudimentary nonreducible building blocks, that of quadsitrons and that of energy. These two building blocks arrange to become the subatomic entities of neutrinos, protons, neutrons and electrons. The subatomic elements create atoms. Atoms bond together to become molecules. Molecules are the construct of the objects which fill the world which surrounds us. The complexity simply becomes greater as the objects become larger.

To understand the principles that comprise the universe requires one to think of the universe as a continuum from the sub^2 atomic fabric to the macromolecules which make up what we see, touch and feel. The molecular structures that combine to create the human form and all the world that surrounds us, organic or inorganic, are constructed from the same sub^2 atomic fabric. All is connected to the to the underlying fabric of space.

The human body is made of up 60% water. Cells, blood, and the space between tissues is generally comprised of water. A water molecule being comprised of one oxygen atom and two hydrogen atoms. We breathe in air which carries water vapor. The atmosphere surrounding us contains water referred to as humidity. We live in a world where we contain water and we are surrounded by water. It is not a difficult leap to conceive that at the subatomic level, the universe is comprised and built from an elemental particle and energy.

The most critical change that needs to occur is the understanding that the universe, which we are able to observe with our eyes, is founded upon a sub^2 atomic universe. This sub^2 atomic universe is comprised of two nonreducible elements, that of energy and that of quadsitrons. Energy is segmented into discrete packets, referred to as quanta. Quadsitrons are four-poled charged units, which represent the smallest unit of matter in the universe. All other components of the universe are derived from these two fundamental elements.

The human experience seeks not just to observe, measure, and mathematically model physical phenomenon like light, but the prepared mind desires a fundamental practical explanation of the means which cause phenomena to exist at all. Curiosity seeks a root explanation at the sub[2] atomic level regarding phenomena such as the nature of the magnetic field surrounding a simple bar magnet, why light and other components of the electromagnetic spectrum propagate as a wave through space, the actual means by which light is reflected, detailed accurate explanation for the unforgiving force of gravity, the mechanics of why a magnetic field comes to exist through and around the earth to protect and nourish life on our planet, and how light is able to travel astronomical distances through space.

Sir Isaac Newton and scientists of his age and before him, pondered that the universe was comprised of a luminous aether.[1,2] The concept of a luminous aether was repudiated in 1887, due to the result of experiments with light, which failed to show the presence of a 'solar wind'. Such a wind effect was proposed to be generated and measurable as the earth plowed through the theorized aether of space as the planet orbits the sun.[3]

With no aether or fabric to comprise the universe, to act as a foundation to theorize physical phenomena, the existence of gravity, light, wave behavior of the electromagnetic spectrum, magnetic fields, and black holes become hollow concepts to be observed, measured and modeled, but not adequately explained at the sub[2] atomic level in such depth as to satisfy human curiosity. Without aether, gravity becomes an unexplained force, the behavior of light becomes a nonsensical phenomenon, the mechanics of the motion of the Milky Way galaxy and other galaxies, becomes no deeper concepts than an artist swathing paint upon a blank canvas.

Physics is best appreciated at the sub[2] atomic level; the world and all of its interactions which exists in the physical realm that is much smaller than an individual proton. The realm micro to a proton is a three-dimensional fluid-like environment comprised solely of energy and elemental particles. The elemental particles identified in this text are quadsitrons. The essential fabric of the universe is an ocean of quadsitrons that expands in every direction to the far fringes of the universe and permeates everything which exists. Quadsitrons and energy are the universe. Everything physical which exists in the universe is a composite of quadsitrons and energy.

The mentioned light experiment of the late 1800's was not capable of taking into account the vast space lying between an electron and the nucleus of an atom. The proposed solar wind could not be accurately measured on the surface of the earth, but rather, as suggested by the theory of relativity, possibly observed and measured at the sub^2 atomic reference window on the surface of an individual proton or neutron, as the atoms comprising the earth, travel through the aether of space.

Physical phenomena including the existence of gravity, light, wave behavior of the electromagnetic spectrum, black holes, and the magnetic field surrounding a bar magnet or the earth are the result of the clash between two Titans, that of nonreducible energy and nonreducible matter. Following 132 years of the luminous aether being set in the shadows of physics, presented here is a mathematical treatise of the mass and diameter of the quadsitron. This proposed elemental nonreducible sub^2 atomic particle, the quadsitron, is depicted to contain four poles, each 120 degrees opposed to each other to facilitate exhibiting four differing states of energy. The quadsitron represents the elemental particle which permeates all space in the sub^2 atomic universe. The time has come to revisit Sir Isaac Newton and his contemporaries' thoughts on a luminous aether comprising the essential fabric of the universe. Students need to utilize this new set point of knowledge to solve the critical physics problems which loom before us, and expeditiously advance our technology to 'gravity field mechanics', before we irreparably destroy the planet with our predecessors' now archaic combustion technology.

SECTION I

CURRENT CONTRADICTIONS IN PHYSICS

CHAPTER 1

130+ YEARS OF CATASTROPHIC DISCONNECTS IN PHYSICS

THE STUDY OF PHYSICS trusts the student to believe in the existence of almost mystical powers which control the universe at large. It is professed that all objects are held firmly to the earth's surface by the force of gravity, yet how this powerful force exerts an effect on the atomic structure of the human body or any object has yet to be clearly defined. There are equations to mathematically represent the magnitude of the force of gravity, but the actual explanation, at the atomic level, for why the apple falls from the tree and lands on the ground remains elusive. The explanation for why two bar magnets attract opposite poles and repel like poles is taught to be the force of magnetism. The definitive means by which a magnetic force exerts itself upon the atoms of an object causing such an object to move from one point in space to another point in space remains undocumented. There is the case of positive and negative charges. Students are taught like charges attract and opposing charges repel, and the neutral state is when positive and negative charges are in balance. Still, the explanation for how a positive charge actually attracts a negative charge, at the functional atomic level, again remains elusive to the student.

the following is a list of a few of the fundamental
contradictions in physics in 2019.

1. What comprises gravity? That is, if one were standing on earth
    and were to drop this book, what is the fundamental action at
    the atomic level which would lead to the book descending and
    striking the floor?
2. E=MC² dictates that light travels at a constant speed, but
    hypothetically the force of gravity can bend light, slow it down,
    and even stop light, trapping light in a black hole.
3. Objects traveling at the speed of light are thought to have no
    mass; how does gravity actually exert a force to bend the path of
    travel of a substance like light that hypothetically has no mass?
4. How the electromagnetic spectrum behaves as sets of waves.
5. How does visible light behave like a wave and in other
    circumstances behave like a ball with mass?
6. How does light travel for billions of light years from distant stars,
    pass through the atmosphere, to be viewed and appreciated by
    stargazers on earth?
7. If substances are made up of only atoms, comprised of protons,
    neutrons and electrons, how actually is light reflected off an
    object to produce color as seen by the human eye?
8. What is magnetism? The physical observable force a bar magnet
    exerts on metal or another bar magnet remains unexplained.
9. The origin of the electromagnetic field which surrounds the
    earth and protects life from fatal solar radiation remains
    unexplained.
10. The mechanics of Beta Decay of a Neutron, has been defined, but
    the properties to properly explain the event remains unexplained.
11. How exactly is a proton constructed?
12. How exactly is a neutron constructed?
13. How exactly is an electron constructed?
14. How is a gold atom, comprised of 79 protons and electrons,
    along with 118 neutrons with numerous orbitals [Xe] $4f^{14}$ $5d^{10}$
    $6s^1$, have a smaller radius (estimated radii 174 pm) than a sodium
    atom (estimated radii 190 pm) with 11 protons and electrons, 12
    neutrons and many fewer orbitals [Ne] $3S^1$?

15. If electrons orbiting the nucleus of an atom are the reason why atoms are perceived to be solid, why don't the many electrons filling the differing shaped and overlapping orbitals contained inside atoms crash into each other?
16. A black hole is thought to be the result of a collapsed star, but if a black hole traps light, then at some point should not a black hole explode due to having accumulated an overabundance of radiant energy?
17. By what subatomic principles does Albert Einstein's observed Photoelectric effect actually work?

What we have been taught is that gravity is an unseen force which causes objects to fall to the earth. Magnetism is an unknown force that causes bar magnets to attach via opposite poles or repel 'like' poles. The mysterious forces which are responsible for gravity and for magnetism certainly exist as practical entities even to the casual observer, but a detailed description of explanation for the 'root cause of every action' of both essential forces remains forthcoming.

In order to appreciate the basis of gravity and magnetism, one must disassociate their consciousness from the laws of conventional physics and open their mind to challenging these paradigms. The stark reality is there are two, and only two components comprising the universe. The two components are energy and quadsitrons. To expound upon this fundamental concept, magnetism and gravity simply represent gradients of energy flow and interaction between energy and quadsitrons. Note, some may say that God is a necessary third component of the universe, but such a discussion is beyond the scope of this text and best left for the theologians to describe and debate.

Quadsitrons act as the fabric of the universe. Light and energy of various wavelengths utilize quadsitrons as a medium to transfer from one location to another location in the universe. Energy flow is a gradient depending upon density (increased energy) and wavelength. Energy flow that transfers along the medium of quadsitrons is low energy. Energy flow which is dense enough to move molecular structures represents a higher tier of energy and is the fundamental concept of magnetism. Gravity represents the highest known tier of energy. Gravity represents such a density of energy that it is capable of physically moving quadsitrons from one location in the universe to another location.

A simple analogy of the gradient of energy is represented by the air that surrounds us and provides a suitable medium to support life on the planet. Still air provides an environment for life to thrive by providing a medium for oxygen utilization. Air which moves, representing wind, produces a disturbance in the environment that is often beneficial for life. Air moving rapidly, at excessive speed, such as a component of a tornado or a hurricane or typhoon, represents the same medium and energy flow, but at a level that is capable of causing disruption of the physical environment leading to damage to the environment. All represent energy flow with the same medium.

The movement of quadsitrons is related to two states. One state is normal flow of quadsitrons created by nonlinear potentials stretching across expanse of the universe. A second state of energy flow is created by trapped energy. Energy is continuously flowing, but may exist in one of two forms: flowing in an endless circuit around a fixed position in the universe, otherwise referred to as 'trapped' or 'dependent' energy; versus 'independent' energy which is flowing from one location to another location without completing a circuit. Energy which is trapped would be represented by the magnetic field of a bar magnet; where energy which is independently flowing from one location to another location would be represent by light being emitted by a distant star and arriving upon earth for a viewer to see and appreciate.

CHAPTER 2

ILLUSIONS OF SOLIDS

AS HUMANS, WE ARE enslaved by the limitations of our senses. The sense of sight, smell, touch, and hot and cold are all meant to allow us to experience the environment within which we live and these senses function exquisitely; but with limitations. The sense of sight allows one to see objects and color, depth perception allows for us to estimate size of objects relative to other objects within our window of vision. The electromagnetic spectrum frequencies encompass energy waves which range from below one hertz to above 10^{25} hertz, and wavelengths of thousands of kilometers in length to sizes small enough to be a fraction of the diameter of the nucleus of an atom. But the energy which a human's eyes generally absorb is a limited portion of the overall electromagnetic spectrum, generally between the frequency range of 430-770 terahertz (1THz = 10^{12} Hz) or wavelength band of 390 to 700 nanometers (1 nm = 10^{-9} m).[4,5] What the right and left halves of the occipital cortex of the human brain is able to appreciate about the environment humans live in, falls within this narrow brand of electromagnetic energy.

The limitations of our sense of touch is even more devastatingly restrictive to a human's understanding of the universe which surrounds us. Yes, if an individual reaches out their hand and touches an object set in front of them, there is an abundance of signals which travel up to the brain. The tip of a finger contains numerous sensors to signal the human brain including tactile, pressure including wet and dry, proprioception or sense of position of a limb, fast/slow pain, hot/cold sense, and vibration sense. See Figure 1. Electrical stimuli travel from the receptors scattered across the hand, up the peripheral nervous system embedded in the arm, jumps onto the nerve tracts of the central nervous system comprising

5

the spinal cord, then continues through the brainstem at the base of the skull, to the parietal lobes of the brain for interpretation. The signals generated by the hand to inform the brain of the characteristics of the object the hand is interacting with generally include a sense of pressure, sense of smooth versus rough surface contour of the object, sense of hot and cold, sense of pain if appropriate, sense of position of the object in space in relation to the orientation of the human body touching the object.

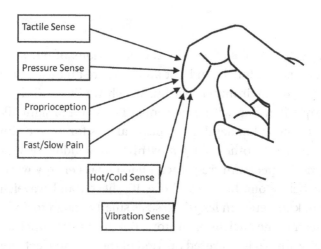

Figure 1

The senses present at the tip of a finger include tactile, pressure, proprioception, fast/slow pain, hot/cold sense, and vibration sense.

Our human sense of touch is a brilliant orchestra of signals, which allows a person to experience and interpret the environment surrounding us. These human senses of touch generally produce a definitive sense of boundaries of objects. We distinguish substances in the environment around us as existing as either solid, liquid or gas. Out of necessity, we classify solid objects of having definitive substance, which generally provides that definitive substance the tangible capability of interacting mechanically with other definitive substances. We tend to classify solids as objects with defined boundaries and a certain mass. Yet, we are intuitively aware, that given enough heat over a sufficient interval of time, any solid object can be transformed into a liquid; most readers have seen molten rock bursting forth from the depths of an active volcano. Further, we are generally aware that again, given enough heat, any solid can be transformed into a gas. The physical properties of water, given the environment of the earth, allow humans to witness ice melting into water, which can be heated to a temperature of 212 degrees Fahrenheit or 100 degrees Celsius to create vapor.[6] Given the proper temperature input, all other solids can be caused to follow a similar physical transformation from solid to gas.

So, we know a solid can become a gas. Generally, a gas, unless contained, has no defined boundaries. The human brain appreciates a gas as having no physical boundaries, but a solid form of the same substance as having definitive boundaries. The perception that a solid has definitive boundaries is rooted in the limitations of our human senses.

All solids are made of up molecules. All molecules are comprised of atoms. All atoms are made of a nucleus and one or more electrons. The hydrogen atom is comprised of one proton and one electron. All other atoms comprising the Periodic Table are generally comprised of protons and neutrons forming a nucleus and electrons in orbit about the nucleus. The number of protons present in the nucleus of an atom define the element. Generally, there are a similar number of neutrons and electrons present in an atom to match the number of protons in the nucleus of the atoms. Isotopes of an element refer to atoms which have a differing number of neutrons or electrons than the element's number of protons present in the nucleus of the atom.

The concept of an atom has generally been described as being likened to a solid. The concept that an atom is a solid has been defended by the statement that the electron circulating the nucleus of the atoms moves at the speed of light, which represents such a high velocity, that for all practical purposes, the electron is in all places in orbit at once and, therefore, an electron orbiting the nucleus of an atom creates a physical, definitive otherwise solid shell around the nucleus of the atom the electron orbits. The shell the electrons create has been thought to generate the solid nature of an atom, and, therefore, the solid nature of substances.

The concept that an electron creates a solid shell by virtue that the electron is moving so fast that the electron is in all places at all times, breaks down in theory when considering that various shells have differing shapes. Atoms are considered to have various shells, which depending upon the number of protons and electrons an element is comprised of may include an 's' shell, a 'p' shell, a 'd' shell, and a 'f' shell.[7] There are also differing layers and progressively larger layers to include a 1 's' shell, then a 2 's' shell and a 2 'p' shell, then a 3 's' shell, a 3 'p' shell and a 3 'd' shell, then a 4 's' shell, a 4 'p' shell, a 4 'd' shell, and a 4 'f' shell and so on. The shells are shaped differently. The 's' shell is in the shape of a sphere, the 'p' shell is in the shape of a dumbbell, the 'd' shell in the approximate shape of a pear, the 'f' shell in the likened shape of a cluster of eggs. The path of the electrons of the various shaped shells theoretically traverse through the center of the atom. If an electron in an orbital is likened to a solid, this becomes difficult to rationalize given that all of the electrons passing through the center of the atom do so without collision. Electrons crashing into each other with in the crowded confines of an atom's densely packed nucleus, should result in the frequent and spontaneous destabilization of atomic structures.

A hydrogen atom is comprised of a single proton at the center of the atom and a single electron at the perimeter of the atom, with the electron orbiting the nucleus of the atom. At the sub^2 atomic level, the distance between the electron orbital and the proton in a hydrogen atom is likened to the distance between the earth and the sun in our macro universe that we can see. In reality, there is significant volume of space contained inside the boundaries of an atom that is not comprised of the protons, neutrons, or electrons. It has been calculated that 99.999999% of an atom represents empty space.[8]

Data collected by human senses are interpreted by the human brain as the touch of an object existing in a solid physical state as being solid. Reality is there is a significant amount of empty space between the nucleus of each atom and the electrons which orbit the nucleus. The objects which fill the environment that surrounds us, at the sub^2 atomic level, are all harbingers of a vast space, this space filled by an ocean of quadsitrons. This ocean of quadsitrons permeates everything at the sub^2 atomic level and extends in all directions to the very boundaries of the universe.

Calculating the properties of the universe, by re-setting one's perception of the universe at the sub^2 atomic level allows for phenomena such as gravity, the behavior of light, magnetism, the existence of black holes, and wormholes to have practical and measurable meaning. Saying that gravity and magnetism exist and demonstrating the effect of both forces on objects does not explain the root cause of the existence of these forces. Gravity, magnetism, and light will be shown to be products of the powerful sub^2 atomic universe.

CHAPTER 3

FATEFUL MISDIRECTION OF 1887

PRIOR TO 1887, MANY of the great minds of science had considered that the universe consisted of an aether. This aether was thought to be the underlying fabric of the world that surrounds us. Sir Isaac Newton described the concept of a 'luminous aether' in his text Optics. Newton felt that light required a transfer medium to travel from one location in the universe to another location in the universe.

In 1887, Albert A. Michelson and Edward W. Morley decided to test the theory of the existence of a luminous aether.[3] The basis of the hypothesis of the experiment was that the earth, the third planet in orbit around the sun, was considered to be a massive celestial object plowing its way through space. It was speculated that if a luminous aether existed, then the earth would create a wake as it passed through such an aether. See Figure 2. The earth travels at a velocity of approximately 67,000 miles per hour in its orbit around the sun.[9] It was hypothesized that if light did depend upon a luminous aether to travel from one location to another location in the universe, then the travel of light should be noticeably affected on the surface of the planet by the wake created as the earth orbited the sun.

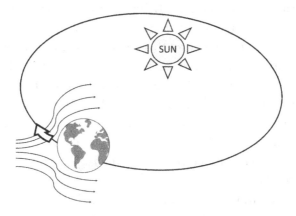

Figure 2

Orbit of the Earth creates a wake in the luminous aether.

In 1887, Michelson and Morley conducted their experiment.[3] They set up two sources of light and positioned the beams 90 degrees in relation to each other. When they activated the lights simultaneously, they measured the time for each beam to make contact with the respective light sensor. See Figure 3. The results of the experiment showed that both beams of light reached their respective target at the identical time. Since there was no difference in the time to target of the emitted light, it was deduced that no luminous aether existed. This belief that there exists no aether has stood for the last 13 decades.

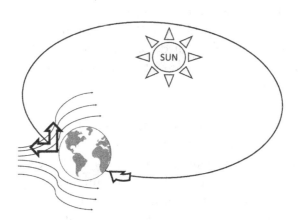

Figure 3

Experiment to test the hypothesis of luminous aether.

The question is not whether Michelson and Morley's measurements were correct or not. The measurements were more assuredly correct. What comes into question was the perspective from which the final analysis of the experiment's findings was derived to arrive at the published conclusion. Michelson and Morley were hampered by the fledgling wisdom regarding atomic chemistry of their time. In 1772, Daniel Rutherford identified nitrogen as one of the four major constituents of the atmosphere and called it mephitic air due to its noxious smell.[10,11] Then in 1783, Cavendish determined the concentration of atmospheric nitrogen to be 79.16%.[12]

Still, understanding that nitrogen existed versus what nitrogen consists of, are two widely differing concepts. Sir Joseph John Thompson an English physicist, whom discovered the electron in 1887, fostered what became a popular 'plum pudding' model of the atom in 1904, describing the atom as 'the atoms of the elements consist of a number of negatively electrified corpuscles enclosed in a sphere of uniform positive electrification'.[13] In 1911, Ernest Rutherford, a New Zealand born physicist, described the atom as containing a positive charge Ne at its centre, and surrounded by a distribution of negative electricity Ne uniformly distributed within a sphere of radius R'.[14] The Bohr model of the atom was introduced in 1913 by Niels Bohr.[15] The Bohr model identified an atom as an entity comprised of a center with electrons circling around the center of the atom in orbitals much like planets orbiting the sun. Bohr also incorporated early quantum mechanics into his model. Michelson and Morley's work in 1887, did not have the advantage of the knowledge of the Rutherford atom model or Bohr atom model or other later models of the atom. The current Electron Shell Model of the atom has been developed over time as a refinement of the Rutherford-Bohr Model.[16]

To satisfy the human senses, the contrived theory is that the electron, moving at the speed of light, in its orbit around the nucleus of the atom, is in all places at once. Since the electron is moving so fast in its orbital around the nucleus, the electron orbital takes on the quality of being a solid. See Figure 4. This theory might be applicable to the Bohr-Rutherford model where electron orbitals were suspected to be concentric spheres. The more modern theory of the electron orbital with irregularly shaped orbitals all seemingly converging on the center of the atom, seems to contradict the solid nature of the electron orbital.

The exact construct of the atom still remains somewhat of a mystery. An atom seems to be much more hallow than solid.

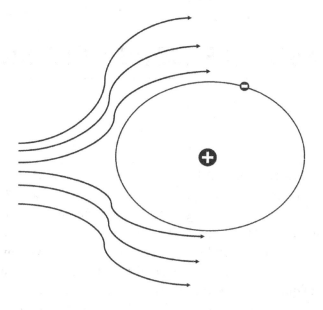

Figure 4

Limitation of human senses approximate an atom existing as a solid.

To be clear, the experiment Michelson and Morley carried out yielded accurate measurements. The conclusions that were made at the time of the experiment were correct given the understanding of the atomic physics in 1887. Given the much more advanced and detailed repository of scientific knowledge that has been accumulated since the late 1800's, we need to revisit the details of the results of the experiment to formulate an updated conclusion.

CHAPTER 4

CRASH CAR DERBY PHYSICS

CURRENT TEACHINGS OF PHYSICS suggest that a proton is comprised of a number of differing particles. There are identified at least 17 possible sub-proton particles, including at least six different relatively well-known units referred to as quarks. See Figure 5. The concept of a quark was devised independently by Murray Gell-Mann and Gnome Zweig in 1964.[17,18,19,20] Today's classic picture of a proton is a subatomic body comprised of quarks. There are a number of identified quarks to include Up and Down quarks, Charm quarks, Strange quarks, Top quarks and Bottom quarks. There are also a number of other sub atomic particles that have been identified.[21]

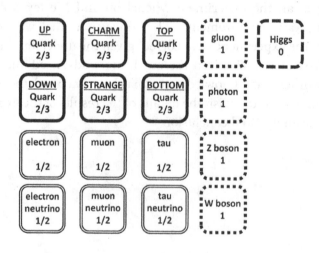

Figure 5

Various sub-units of a proton.

Protons are not usually stationary when they exist outside the confines of the center of an atom. It is difficult to accurately study any object that is constantly in motion. Therefore, the study of the construct of protons is generally been done in the context of what happens to a proton when it collides with another subatomic particle. Thus, the general knowledge of a proton is related to how a proton breaks apart after a collision. See Figure 6.

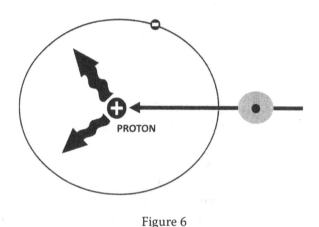

Figure 6

Proton hit by a subatomic particle to break the proton apart.

The pieces which result following a proton colliding with another subatomic particle are not necessarily an accurate representation of the construct of the two original objects. An analogy would be if one were sitting in the stands viewing a car crash derby: cars participating in the derby were covered such that the make and model of the cars were kept secret from the crowd and the only means of determining how the vehicles participating in the derby were constructed, were by piecing together the parts of each vehicle which fell off following collisions between vehicles. See Figure 7. It would be difficult to visualize how a vehicle was constructed if one's study of the vehicle was restricted to access only to the pieces of the vehicle that randomly fell off following each collision. Not only would the vehicle be in pieces and reconstructing the architecture of the vehicle would be similar to piecing together a puzzle without a reference point, but the pieces one would be using would be damaged pieces due to forces of the collision, and thus not necessarily representative of the vehicle prior to collision.

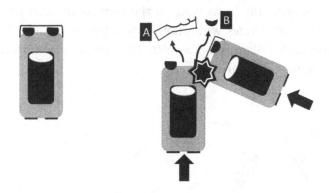

Figure 7

Crash car derby physics analysis.

Likely there is an essential building block that is utilized to build a proton, a neutron and an electron. But smashing subatomic particles racing at high speed into a proton to determine the contents of a proton would yield only limited useful information regarding the true character of a proton. See figure 8.

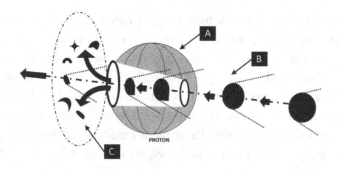

Figure 8

Planned collision of a proton with a high-speed subatomic particle resulting in pieces being ejected from the body of the proton. In the illustration 'A' represents a proton, 'B' represents a high-speed atomic particle, 'C' represents random debris post collision.

When a proton is struck by a high-speed subatomic particle which pierces the proton, the proton ejects a portion of its structure into the space surrounding the proton. The fact that the proton is reducible to smaller pieces opens the door to theories that the universe is comprised of a uniform particle smaller in size and weight than a proton. Trying to decipher the exact dimensions of the nonreducible sub proton particle by colliding subatomic particles into protons is more than likely not going yield an optimal description of the nonreducible sub proton particle.

CHAPTER 5

REALITY: THERE IS NO POSITIVE OR NEGATIVE CHARGE

THERE IS NO POSITIVE or negative charge as most of us have been led to believe. That is, like the forces of gravity and magnetism, students have been led to believe there is are invisible forces of positive and negative that somehow mystically exist in the universe. The proposed forces of positive and negative have been described as tending to be opposing like forces and somehow attracting unlike forces to each other. The invisible forces of gravity, magnetism, positive and negative are at best myths. Gravity, magnetism, positive and negative forces have all been contrived as almost transcendent entities to explain to our brain what our senses tell us about the environment which surrounds us. The details of how gravity, magnetism, positive and negative really work at the level of atomic structures is lacking. A more realistic approach to understanding the function of these forces is the concept that the universe is only comprised of energy and quadsitrons and every other phenomenon which exists is constructed from these two fundamental entities.

Instead of a positive or negative charge or pole for an object, there should alternatively and more correctly be an identifying label to indicate a flow of energy into a node or flow of energy out of a node. Energy is in constant motion. Energy may be free, moving from one point in the universe to another point in the universe, or energy may be trapped. Trapped energy is corralled into a loop. Therefore, the energy continues to cover the same territory, but remains in constant motion. Energy may exist as a wave with a frequency, or energy may exist without a wave pattern and simply exist as a flow.

Energy has density. A quanta is the fundamental unit of energy. Similar to a river of water being comprised of many raindrops or a single raindrop being comprised of numerous water molecules, an energy flow may be comprised of numerous quanta, as described in quantum mechanics. The density of the flow of energy defines whether the energy represents a surge of the electromagnetic spectrum, magnetism or gravity.

A pulse of the electromagnetic spectrum propagates as a wave from one point in the universe to another point in the universe. The wavelength of the wave determines the type of electromagnetic pulse the wave represents. Again, wavelength of 390-700 nm represents the spectrum of visible light.[4] The waves of the electromagnetic spectrum require an aether comprised of quadsitrons as their transfer medium, to transit from one location of the universe to another location of the universe.

Magnetism represents energy denser than the electromagnetic spectrum. Magnetism represents a form of trapped or channeled energy. Quadsitrons line up in loops trapping energy that passes through an object which exhibits magnetic field properties. Such an object is usually referred to as a magnet. The density of magnetic fields is such that they can move objects comprised of molecules.

Gravity is the densest form of the three forms of energy. Gravity reaches a density of energy, that it is capable of manipulating the flow of quadsitrons. Manipulating or diverting the flow of quadsitrons results in altering the physical shape and thus the dynamics of the universe.

There is no mysterious positive pole or negative pole to a physical object. Magnets do not have a positive pole or a negative pole. There is only flow of energy out of an object and flow of energy into an object. Positive might arbitrarily represent flow of energy out of an object and negative might arbitrarily represent the flow of energy into an object.

A bar magnet is a good example, where by convention there is generally designated a 'positive' or north pole of such a magnet and a 'negative' or south pole of such a magnet. A bar magnet, by the atomic structure of the magnet, creates channels in the quadsitrons located inside the physical boundaries of the magnet and quadsitrons adjacent to the bar magnet. The channels of quadsitrons inside the bar magnet and surrounding the bar magnet trap energy that the bar magnet comes in contact with, and causes the energy to circulate through the channels until that point where the material properties of the bar magnet are disrupted or the energy is drawn out of the looping channels by a more intense sink or attracter of energy. There are numerous channels associated with any bar magnet. The more channels the atomic structure of a bar magnet is capable of supporting, the more quanta the magnet is capable of trapping, the denser the energy flow through and around the magnet, and thus the stronger the magnetic force of the bar magnet.

SECTION II

DEFINING AN ELEMENTAL PARTICLE

CHAPTER 6

RE-CONSTRUCTING PHYSICS WITH AN AETHER

AS SIR ISAAC NEWTON promoted, a luminous ether likely exists constituting the underlying fabric of the universe. As noted in Newton's text *Opticks* 'And so if anyone should suppose that Aether (like our Air) may contain Particles which endeavor to recede from one another (I do not know what this Aether is) and that its Particles are exceedingly smaller than those of Air...'.[1]

Prior to 1887, the popular scientific belief was that a luminous Aether filled the universe.[2] Sir Isaac Newton promoted the concept of the luminous ether, where he referred to light utilizing a medium for transfer.[1] Light was thought to utilize the luminous ether to travel any distance from one location in the universe to another location.

In honor of the insightful scientists of Sir Isaac Newton's age and earlier, who believed in the existence of an ether, this text will use the European form of the word ether, which is 'aether'.

In 1887, Albert A. Michelson and Edward W. Morley discredited the idea a luminous ether filled the space between heavenly bodies.[3] Michelson and Morley theorized that if a luminous ether existed, then the earth would generate a solar wind as the mass of the planet passed through such an aether in space while the planet orbited the sun. Michelson and Morley theorized that if such a solar wind existed on the surface of the planet, then this solar wind would disrupt the luminous aether resulting in a measurable effect on the speed of light on the surface of the planet. Michelson and Morley conducted an experiment in which they measured the speed of light of two beams of light perpendicular to each other, and found there was no difference in the speed of light of the two beams of light. Michelson and Morley, their contemporaries, and those whom have followed concluded that since there was no difference in the speed of light between the two beams of light perpendicular to each other, no luminous aether existed.

If space truly consists of a vacuum, generally void of any substance, it would seem the elements consisting of matter, would have been migrating towards some common state of least potential for the last 13.77 billion years, the estimated life of the universe.[22] A similar situation would be illustrated by a thunder cloud raining down on the side of a mountain, for the most part, the water comprising the rain would strike the side of the mountain and any excess moisture would make its way towards the point of least potential at the bottom of the mountain and form a brook, then a creek, then a river and finally flow into a lake.

It is understood that space is not completely empty. Stars emit various forms of energy seen in the electromagnetic spectrum, often including gamma rays, x-rays, visible light, and radio waves. Our sun, a yellow medium star, emits most of its energy as visible light, infrared light and heat.[23] Stars also emit gases and plasma, electrically conductive ionized gas. We know space is occupied by dust, debris, asteroids, comets, planets, moons, and stars. Space, therefore, is not a true vacuum.

The Big Bang Theory has been the dominant theory to describe the origin of the universe.[24,25] The position taken by the Big Band Theory argues against the existence of a luminous aether. The Big Bang Theory refers to the concept that the universe started as a single concentrated mass of matter, then following the build up a critical amount of energy, a colossal explosion occurred. The explosion marking the beginning of time, for the current cycle, flung mass in every direction, creating what we know to be the universe today. The theory continues, exerting the concept the universe is either continuously expanding or contracting, including the idea that following the initial Big Bang explosion, matter was flung toward the distant reaches of the universe, finally to reach a point where the tide reverses and the movement of matter in the universe reverses from expansion to contraction. The Big Bang Theory teaches that the universe has been cycling between consisting as a single point of matter, to a large expanse of stars, and back to a single point of matter. The current age of the universe has been estimated to be 13.7

billion years old.[22,26,27] The age of the earth is thought to be 4.54 billion years.[28,29,30,31]

If the universe was in truth a void comprised of no aether or other form of consistent fabric, then this sets up a physical paradox. If there is no aether underpinning the elements of the universe then everything in the universe should gravitate to the lowest potential. The Big Bang Theory would suggest all matter is either traversing out from one point or retreating toward one point in space, that is the origin of the big bang. It is understood that the earth's Sun resides in the Orion-Cygnus Arm of the Milky Way Galaxy.[32,33] The Sun and the other 100-400 billion stars comprising the Milky Way Galaxy swirl in a circle around the center of the galaxy.[34,35,36] Is it also understood that galaxies traverse the universe and even collide with each other. The Andromeda Galaxy, nearest galaxy to the Milky Way Galaxy, is 2.5 million light years from Earth.[37] If the Big Bang Theory is true, then all matter should be jettisoned from a single point, all matter should have the same inertia and similar trajectory outward from the origin of the big bang, and no matter should be crossing paths with other matter; that is, galaxies should not collide.

It is difficult to conceptualize that if the phase of the universe, according to the Big Bang Theory, is expansion, it is hard to believe that as all of the matter of the universe is moving away from the center of a colossal explosion, that it would create spiral galaxies. In addition, if all matter is moving away from a central point with the same inertia and outward direction, then how do such galaxies follow transverse pathways, and collide into each other.

The radius of an electron per CODATA 2014 is $2.8179403227 * 10^{-15}$ m. CODATA is a repository of standards which can be accessed on the internet.[38] The diameter of an electron is therefore $5.6358806454 * 10^{-15}$ m. The value of Bohr's radius per CODATA 2014 is $5.2917721067 * 10^{-11}$ m.[38] Therefore, the number of electrons that can fit side by side along the radius of a hydrogen atom is calculated to be 18,778. The calculated diameter of a quadsitron is $8.557941465 * 10^{-19}$ m (Eq 79). The number of quadsitrons that could fit side by side along the radius of a hydrogen atom is calculated to be 61,834,637.

The diameter of the earth is $1.2742 * 10^7$ m.[39] The distance from the sun to the earth is $1.496 * 10^{11}$ m.[40] The number of earth-like planets that could fit in the distance from the sun to the earth can be calculated to be 11,740.

More electrons would fit between the proton and the electron in a hydrogen atom than the number of earths that would fit between the earth and the sun in our solar system. From the perspective of the proton, there exists considerable empty space between the proton and the electron in a hydrogen atom. Rutherford, himself, described the atom as being comprised of mostly empty space.[41]

CHAPTER 7

ERROR OF THREE CHARGE STATES

IT HAS BEEN TAUGHT that there exist three states of physics to include 'positive', 'negative' and 'neutral'. It is often taught that the charge states of 'positive' and 'negative' sum to equal 'neutral'. The matter of a positive charge combining with a negative charge has generally considered to neutralize each other and the result has been termed a neutral charge or a neutral pole. Thus, the universe has been thought to exist as three possible states: positive, negative or neutral.

The study of physics suggests the universe is comprised of a proton, electron and a neutron. The proton is by convention, considered to exhibit a hypothetical 'positive' charge. The electron is, again by convention, considered to exhibit a hypothetical 'negative' charge. The neutron is considered, by convention, to exhibit a hypothetical 'neutral' charge.

The problem is that atomic physics does not behave in a manner that dictates that the universe exists in only three states. Conventional wisdom clashes with reality, when the details of neutron Beta decay are studied. Beta decayed is a term applied to the process of a neutron transforming into a proton. A neutron is larger and heavier than a proton.

When a neutron undergoes the process of Beta decay and converts to a proton, the result also produces an electron and a measure of energy identified as a neutrino.[42] The physics of a neutron undergoing Beta decay identifies that in fact a neutron does not only equal the sum of a proton and an electron, but also contains a charge state equating to more than the sum of a proton and an electron. If a neutron represents the charge state of 'neutral' and a proton represents the charge state of 'positive' and an electron represents the charge state of 'negative', in effect, a neutron does not exactly represent precisely or correctly the sum of a proton and an electron, therefore the charge state of neutral does not represent the sum of the charge states of 'positive' and 'negative'. Neutral represents its own charge state. See Figure 9.

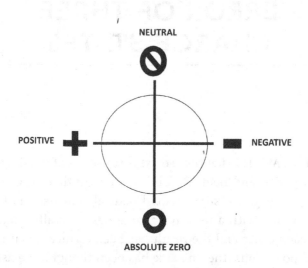

Figure 9

Four states of charge: Positive, Negative, Neutral and Absolute Zero.

Given that the Beta decay of a neutron demonstrates that a neutron has its own distinct value, which is greater than the sum value of a 'positive' combined with a 'negative', then a neutron must be considered to represent its own electric charge state which is detached from the concept that a neutron represents the value of zero. It must also be considered, that the state of 'neutral' is also detached from the value of zero. In effect, zero is its own charge state, separate from a neutron and separate from the term neutral. For purposes of this text, Zero, in effect is termed and referred to as 'absolute zero' to avoid any future confusion.

Thus, there are four states of charge in the universe. The four states of charge are identified as 'positive', 'negative', 'neutral', and 'absolute zero'. Each state of charge is separate from the other charge states and represents a differing value.

CHAPTER 8

QUADSITRON: THE UNIVERSE'S ELEMENTAL PARTICLE

THEORY BEHIND THE PARTICLE

IF ONE ACCEPTS THE fact that there is an underlying fabric to the universe, which Sir Isaac Newton referred to as the luminous aether, then this fabric must be comprised of some elemental particle or entity. Standing on the shores of New England, or the Gulf of Mexico, or on the west coast peering over the ocean as the sun sets, the cubic miles of ocean water which spread out towards the distant horizon are all comprised of individual water molecules, which combine to make what the human senses detect as water. The fabric of the universe, which is a three-dimensional ocean comprised of some extremely minute entity, permeates everything and extends outward to the distant edges of the universe in every possible direction.

This fabric must be comprised of some theoretical entity, some imaginable particle, some discernable concept. Such an entity must be of a construct that insures that when such particles are collected together in a tight space they do not lock up or become amassed. Such particles must never combine together and create a solid form. They must resist conjoining. The surface of such an entity must therefore be some form a sphere, with smooth edges such that the surface of one particle smoothly slips across the surface of an adjacent, similar particle. But simply being spherical, with smooth edges, would not insure that given enough outside pressure, that a group of such particles would not lock up and form a tight ball or other formation. If light uses such essential, nonreducible particles as its transfer medium through space, then such particles must be always be free to move independent of each other and independent of the surrounding forces throughout the universe, no matter how powerful such forces might be in a given location.

The essential particle in question must be extremely small in size and weight. Such an entity must be so small that it permeates everything in the universe. Such a partible must be like water, where it fills the entire glass. Such a particle must be present everywhere in the universe, similar to swimming at the beach and sand gets everywhere including inside one's swim suite and in one's hair. Such an entity must be so small that it does not appreciably impede the physical movement of protons and neutrons as atomic structures passing through space, similar to a scuba diver swimming unimpeded underwater.

To accomplish such a design as to create an entity which is always able to move, no matter how strong an outside pressure or force is raining down on such a particle or group of particles, this entity must be comprised of both matter and energy. A shield of pure energy must surround each essential particle in totality. Where wheels are attached to the axle of a motor vehicle, metal ball bearings are generally packed in grease. The presence of lubricant in a ball bearing joint is meant to reduce friction and facilitate slippage of the steel balls within the joint. Energy, trapped by and surrounding the exterior of an essential particle, would facilitate that such essential particles would never come in direct contact with each other; the surfaces of such essential particles would never make contact. Given that such particles are capable of trapping energy and interacting with energy, the more pressure applied to a group of essential particles, would paradoxically strengthen the individual energy field around each particle and prevent contact of such particles, no matter the magnitude of the exterior forces applied to a group of essential particles. In this text, the theoretical essential particle has been given the name of 'quadsitron'.

DERIVING THE FEATURES OF A QUADSITRON

Traditionally, there have been three recognized states of charge to include positive, negative, and neutral. Neutral has generally been synonymous with nullifying total charge when combining an equal amount of positive charge with an equal amount of a negative charge; Neutral has often been synonymous with the value of zero. Analysis of Beta decay of a free neutron to a proton suggests that there are four states of charge to include: positive, negative, neutral, and absolute zero. Neutral is not representative of zero in the context of Beta decay of a free neutron (neutral) to a proton (positive) and an electron (negative). Neutral, as reference in the realm of physics, is its own independent state of charge in relation to positive, negative, and zero.

Quadsitrons represent a four pole nonreducible sub-sub atomic elemental particle. All quadsitrons are physically identical, including size, shape, weight, and charges expressed on the surface of the particle. A quadsitron contains four differing poles, one of positive, one negative, one neutral, and one absolute zero. Each of the four poles of a quadsitron project 120 degrees from the neighboring poles as depicted in Figure 10. The actual angle between the poles may be less than 120 degrees due to thickness or width of the poles. Methane is a molecule comprised of one carbon atom surrounded by four hydrogen atoms, which is likened to the shape of a quadsitron. The angle between the hydrogen atoms surrounding the carbon atom at the center of the methane molecule is 109.5 degrees due to the size of the covalent bonds. For purposes of this discussion we will assume the angle between poles for a quadsitrons is 120 degrees until further information is available.

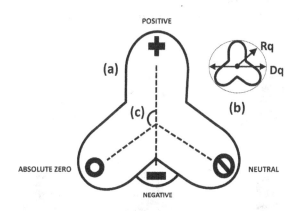

Figure 10

Conceptual illustration of a quadsitron. (a) Four poles comprising a quadsitron. (b) Rq = radius of quadsitron, Dq = diameter of quadsitron. (c) Angle 120 degrees between each of the four poles.

Quadsitrons comprise the very fabric of the universe permeating all known space, analogous to water filling a glass. Various differing configurations of quadsitrons result in the construct of neutrinos, neutrons, protons, electrons, and dark matter.

SUMMARY OF CALCULATIONS OF WEIGHT AND DIAMETER OF QUADSITRONS

The weight of a single quadsitron is approximated to be $9.5075301706 * 10^{-38}$ kg Eq (20)(see Appendix). The diameter of a single quadsitron is calculated per volume of a neutron (Vn) and volume of a proton (Vp) Eq (79) equal to $8.557941465 * 10^{-19}$ m.

The number of quadsitrons calculated by volume and calculated by weight for a neutron are both equal at $1.76168 * 10^{10}$ (Eq (108)). The number of quadsitrons calculated by volume and calculated by weight for a proton are both equal at $1.75926 * 10^{10}$ Eq (113). The number of quadsitrons calculated by volume and calculated by weight for the change of a neutron to a proton are both approximately equal at $2.42499 * 10^7$ Eq (127). The change in the radius of a neutron to a proton is $5.11104551 * 10^{-19}$ m Eq (131). The number of quadsitrons comprising the radius of a proton is 1,300.409179 or truncated to 1,300 Eq (137). The number of quadsitrons comprising the radius of a neutron is 1,301 Eq (143).

Differing angles of view of the quadsitron appear in Figures 11 and 12.

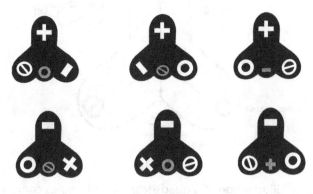

Figure 11

Positive on top in the top row and negative on top in the bottom row.

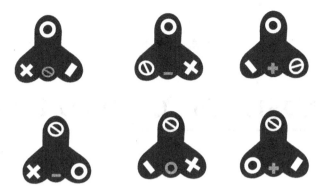

Figure 12

Absolute zero on top in the top row and neutral on top in the bottom row.

To conceptualize a three-dimensional view of the quadsitrons, Figure 13 is provided. The illustrations in Figure 13 are that of a quadsitron tumbling through space. In Figure 13, on the left-hand side of the illustration, the neutral pole is on top with the negative pole on the left-hand side, absolute zero in the middle, and the positive pole on the right-hand side of the quadsitron.

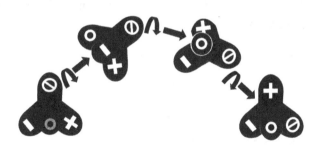

Figure 13

Illustration of a quadsitron tumbling through space
to demonstrate three-dimensional features.

The quadsitron represents the elemental particle comprising the fabric of the universe. A quadsitron is comprised of four differing charged poles. Each charged pole representing positive, negative, neutral, and absolute neutral are physically 120 degrees from the other charged poles.

CHAPTER 9

QUADSITRON Q-FIELD LINES

THE QUADSITRONS EXHIBIT TWO energy fields representing linear flow of energy positioned ninety degrees to each other. The flow of energy in a straight line with no wave pattern or energy outside the visible light spectrum is energy that cannot be detected by the human eye. Energy consistently flows from the hypothetical positive pole into the hypothetical negative pole; in addition, energy consistently flows from the hypothetical neutral pole and into the hypothetical absolute zero pole. When the flow of energy is dense enough, energy can alter the flow of quadsitrons. Energy flowing at ninety degrees to a separate additional flow of energy, does not interfere with the second flow of energy. See Figure 14.

Figure 14

Quadsitron with Q-field lines. Arrows illustrate flow of energy; two flows of energy are ninety degrees out of phase with each other.

Q-field lines represent the flow of energy from the positive pole of a quadsitron to the negative pole and the flow of energy from the neutral pole of a quadsitron to the absolute zero pole of the quadsitron. Figure

15 demonstrates the picture of the quadsitron with Q-field lines as the quadsitron tumbles through space.

Figure 15

The quadsitron with Q-field lines tumbling through space.

The quadsitron has Q-field lines. The Q-field lines create a miniscule margin of separation between each of the quadsitrons. The Q-field lines create a behavior of the quadsitrons that they slip across each other, preventing the quadsitrons from being compressed together and locking together. The Q-field lines also facilitate alignment of the quadsitrons to create neutrinos, magnetic field lines, protons, neutrons and electrons.

The energy of quadsitron field lines can be calculated using Albert Einstein's equation $E = MC^2$. In this case the M or mass is equal to that of one quadsitron. Calculating the mass of one quadsitron equals $9.5075301706 * 10^{-38}$ kg Eq (20)(see Appendix). C2 or light speed squared is equal to $8.98755179 * 10^{16}$ m^2/s^2. The product of the two numbers equals $0.854494198 * 10^{-20}$ joules. There are two field lines, each 90 degrees out of phase with each other, in the construct of a quadsitron. Each field line equates to $0.427247099 * 10^{-20}$ joules.

CHAPTER 10

RIGHT AND LEFT HANDED QUADSITRONS

THE ASTUTE READER, WHO would model the quadsitron in clay to observe the three-dimensional structure of this elemental particle, would quickly realize that the arrangement of the four poles of the particle could result in two differing structures. Therefore, the universe is comprised of a four-poled quadsitron that for illustrative purposes, is either right-handed or left-handed. A right-handed quadsitron (R-quadsitron), in three-dimensional space, is defined as having its positive pole on right with the negative pole on the left, the absolute zero pole pointed upwards and the neutral pole pointed toward the back. See Figure 16.

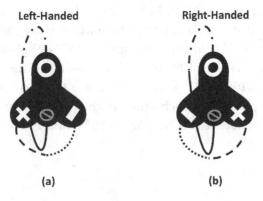

Figure 16

(A) A left-handed quadsitron in three-dimensional space, is defined as having its positive pole on the left with negative pole on the right, the absolute zero pole pointed upwards and the neutral pole pointed toward the back. (B) A right-handed quadsitron has its positive pole on right with negative pole on the left, absolute zero pole pointed upwards and the neutral pole pointed toward the back.

Different perspectives of left-handed quadsitrons (L-quadsitron) and right-handed quadsitrons (R-quadsitron) are presented in Figure 17.

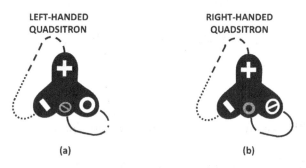

Figure 17

(a) Left-handed quadsitron has its positive pole on top with negative pole on the lower left, absolute zero pole pointed lower right and the neutral pole pointed toward the back. (b) Right-handed quadsitron has its positive pole on top with negative pole on the left, the neutral pole pointed toward the lower right and the absolute zero pole pointed backwards.

Clay was used to model the right-handed and left-handed quadsitrons. Presented in Figure 18, is a clay model of a left-handed quadsitron positioned on the left side of the image and model of a right-handed quadsitron positioned on the right side of the image. The 3D clay models in Figure 18 demonstrate the appearance of the positive pole pointed to the left in the left-handed quadsitron positioned on the left-hand side of the image and the positive pole pointed to the right in the right-handed quadsitron positioned on the right-hand side of the image. The absolute zero pole is pointed upward and to the back, while the neutral pole is pointed downward and to the back in both clay models.

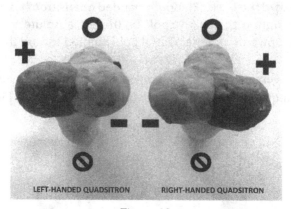

Figure 18

Clay models of the left-handed and right-handed quadsitrons, with absolute zero pole pointed upward and to the back.

Presented in Figure 19, is a clay model of a left-handed quadsitron positioned on the left side of the image and model of a right-handed quadsitron positioned on the right side of the image. The 3D clay models in Figure 19 demonstrate the appearance of the positive pole pointed to the left in the left-handed quadsitron positioned on the left-hand side of the image and the positive pole pointed to the right in the right-handed quadsitron positioned on the right-hand side of the image. The absolute zero pole is pointed upward and toward the viewer, while the neutral pole is pointed backward and away from the viewer in both clay models.

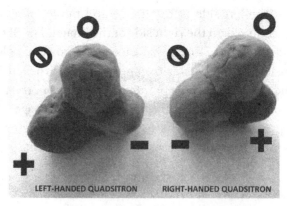

Figure 19

Clay models of the left-handed and right-handed quadsitrons, with absolute zero pole pointed upward and toward the viewer.

There are a number of binding advantages for having right-handed and left-handed quadsitrons. The percentage of R-quadsitrons versus L-quadsitrons in the universe is unknown at this time and may have started out as fifty-percent for each type of quadsitron. The ebb and flow of the universe, may have mixed the percentages of R-quadsitrons versus L-quadsitrons into differing proportions at different locations throughout the universe, which may account for differing phenomena scattered across the cosmos.

Now that the quadsitron has been discussed as a positive, negative, neutral and absolute zero poles, and right handed and left-handed, in reality the universe does not concern itself about such detail. All the universe is concerned about is an elemental particle comprised of four poles with two loops of energy, each energy loop flowing 90 degrees out of phase with each other. The above description of the quadsitron is meant to allow the human brain to analyze and comprehend the existence of the elemental particle. In actuality, the elemental particle is a generic and functional.

CHAPTER 11

NEW PERSPECTIVE APPLIED TO EXPERIMENT OF 1887

MICHELSON AND MORLEY SET up their experiment to investigate for possible evidence of a luminous aether.[3] At the conclusion of collecting their data regarding comparing two light beams positioned at a 90-degree angle from each other and the two light beams reaching their perspective targets at the same time, Michelson and Morley surmised that there must not be a wake, therefore there must not exist a luminous aether. Again, the question is not whether Michelson and Morley's measurements were correct or not, and in addition, whether their interpretation of their experiment's findings were correct or not, for both the conduction of the experiment and the conclusion given the state of atomic physics and molecular chemistry in 1887, were accurate.

The Bohr model of the atom, building on the Rutherford model of the atom, was introduced in 1913 by Niels Bohr.[15] The Bohr model identified an atom as an entity comprised of a center with electrons circling around such a nucleus, and utilized early quantum theory. The current Electron Shell Model of the Atom has been developed over time as a refinement of the Rutherford-Bohr planetary model.

As Cavendish discovered, the atmosphere of earth is comprised of 79% nitrogen. Oxygen makes up nearly 21% of the atmosphere.[12] The remaining portion of the atmosphere is comprised of carbon dioxide, water vapor and numerous other gases.

So, to understand Michelson and Morley's experiment and the outcome that the experiment would have arrived at, one should investigate the components of the medium light travels through the atmosphere.

Investigating nitrogen and oxygen, which combined comprises the vast majority of the atmosphere on the planet's surface, both entities generally exist as molecules comprised of two like atoms. The nitrogen molecule in the atmosphere is comprised of two nitrogen atoms held together with a triple bond. The oxygen molecule, as it generally exists in the atmosphere, is comprised of two oxygen atoms held together with a double bond.

The nucleus of a nitrogen atom contains seven protons and seven neutrons. The nitrogen atom has a calculated atomic radius of 56 pm (picometers) or 56 x 10^{-12} m.[43] The diameter of a nitrogen atom is 112 pm. The atomic mass of nitrogen is 14.00674 amu.[44] A nitrogen atom has seven elections distributed as two in the 1s shell, two in the 2s shell and three in the 2p shell.

The Nitrogen atom has a number of isotopes. The two stable isotopes of nitrogen are ^{14}N and ^{15}N. Of the stable isotopes ^{14}N makes up 99.634% of natural nitrogen.[45,46]

A nitrogen molecule consists of two nitrogen atoms referred to as a dinitrogen, bonded together with a triple bond. Two nitrogen atoms each share three electrons from the 2p orbits to create one of the strongest bonds in nature. The kinetic diameter of a nitrogen molecule is approximately 364 pm.[47]

Earth has an equatorial diameter of 12,742,000 meters.[39] The distance from the earth to the sun is 149,597,870,700 m.[40] The sun has a diameter of 1,392 million km.[48] Given the distance from the earth to the sun is 149,597,870,700 m, then the number of earth like planets which could exist on the radius from the earth to the sun would be calculated to be 11,740.53.

The radius of the nucleus of an atom can be approximated by using the following equation:

$$R = r_0 * A^{1/3},$$

where R is the radius of the atom, A is the atomic mass number (number of protons and neutrons), and r_0 is 1.25 x 10^{-15}. Using this equation, the radius of nitrogen atom can be calculated as 3.012651 fm, where fm equals 10^{-15} m. The diameter of the nucleus of a nitrogen atom is 6.02530 fm.

The radius of an electron per CODATA 2014 is $2.8179403227 * 10^{-15}$ m. The diameter of an electron is therefore $5.6358806454 * 10^{-15}$ m. The value of Bohr's radius per CODATA 2014 is $5.2917721067 * 10^{-11}$ m.[38] Therefore, the number of electrons that can fit side by side along the radius of a hydrogen atom is calculated to be 18,778.

The calculated diameter of a quadsitron is $8.557941465 * 10^{-19}$ m. The number of quadsitrons that could fit side by side along the radius of a hydrogen atom is calculated to be 61,834,637.

More electrons would fit between the proton and the electron in a hydrogen atom than the number of earths that would fit between the earth and the sun in our solar system. Three hundred twenty-nine times the number of quadsitrons versus electrons would fit between the electron and proton in a hydrogen atom. Hydrogen is a lighter gas than nitrogen or oxygen and therefore would tend to float toward the periphery of the atmosphere, the boundary between earth and space.

The majority of earth's atmosphere at the planet's surface is nitrogen, which comprises 79% of the breathable air. The calculated atomic radius of a nitrogen atom is 56×10^{-12} m. The diameter of the nucleus of a nitrogen atom is 6.02530 fm, therefore the radius is 3.012651 fm. The distance between the outer limits of a nitrogen atom and the nucleus of the nucleus is 56×10^{-12} m subtract 3.012651×10^{-15} m, which equates to $55,995.987349 \times 10^{-15}$ m.

As mentioned above, the radius of an electron per CODATA 2014 is $2.8179403227 * 10^{-15}$ m. The diameter of an electron is therefore $5.6358806454 * 10^{-15}$ m.

The diameter of a quadsitron is calculated to be $8.557941465 \ 10^{-19}$ m. Therefore, 65,430,469 quadsitrons can exist along the radius of a nitrogen atom between the outer limit of the nitrogen atom and the nucleus of the nitrogen atom.

Object	Distance	Quantity
Earth diameter	12,742,000 m	---
Sun diameter	1,391,400,000 m	---
Distance Earth to Sun	149,597,870,700 m	# of Earths = 11,741
Electron diameter	5.635880645410^{-15} m	---

Hydrogen Atom nucleus to distant electron	5.291772106710^{-11} m	# of Electrons = 9,389
Nitrogen Atom nucleus to distant electron	77.5 x 10^{-12} m	# of Electrons = 13,751
Kinetic Diameter of Nitrogen Molecule	360 x 10^{-12} m	---
Kinetic Radius of Nitrogen Molecule	180 x 10^{-12} m	# of Electrons = 31,938
Quadsitron diameter	8.557941465 10^{-19} m	---
Hydrogen Atom nucleus to distant electron	5.291772106710^{-11} m	# of Quadsitrons = 61,834,637
Nitrogen Atom nucleus to distant electron	55.995 10^{-12} m	# of Quadsitrons = 65,430,469
Kinetic Radius of Nitrogen Molecule	180 x 10^{-12} m	# of Quadsitrons = 210,330,955

Table 1

Rutherford described the atom as being comprised of mostly empty space.[41] From the perspective of the proton, there exists considerable empty space between the proton and the electron in a hydrogen atom. There exists a vast empty space between the outer electrons and the nucleus of a nitrogen atom. Similarly, there is significant space between the Earth and the Sun in the solar system. See Figure 20.

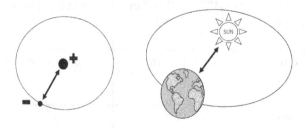

Figure 20

(a) Significant space exists between the electron and the nucleus of an atom,
(b) similar to the significant space which exists
between the Earth and the Sun.

The above calculations suggest that 11,740 earths could exist in a single line between the sun and the earth, and in comparison, over 65 million quadsitrons could exist in a single line between the nucleus of a nitrogen atom and the outer limits of a nitrogen atom. The calculations suggest that in reference to the calculated diameter of a quadsitron and comparing a nitrogen atom to the earth-sun relationship in the solar system, there is proportionately more space between the outer electron rings and nucleus of a nitrogen atom than between the earth and the sun in the solar system.

The solid matter of a nitrogen atom, the volumetric size of the nucleus of a nitrogen atom, comprises a miniscule portion of the actual volume of a nitrogen atom. Given the small size of a quadsitron in comparison to the nucleus of a nitrogen atom and the overall volume of a nitrogen atom, the majority of the flow of quadsitrons that pass through a nitrogen atom at any given time are not affected by the passage of a nitrogen atom from one point in space to another point in space. The atoms comprising the earth float in a three-dimensional sea of quadsitrons.

In its gaseous form, two nitrogen atoms bind with a triple covalent bond. The kinetic diameter of a nitrogen molecule is 370 pm, which is triple the size of the diameter of two individual nitrogen molecules. See Figure 21. The number of protons and neutrons and electrons simply double, but otherwise do not change when adding to nitrogen atoms together to make a nitrogen molecule. Therefore, there is even more open space when considering nitrogen molecule in a gas state, such as would be found in earth's atmosphere.

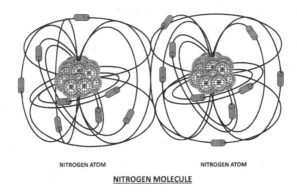

Figure 21

Illustration of a Nitrogen molecule.

As calculated above presented in Table 1, the number of electrons which could fit along the radius of a nitrogen molecule would be 31,938, and the number of quadsitrons which could fit along the radius of a nitrogen molecule would be 210,330,955. There is a significant amount of volume to the atmosphere comprised predominantly of nitrogen gas, which is void of subatomic particles.

Oxygen atoms generally consist of 8 protons and 8 neutrons in the nucleus and 8 electrons in orbit. Oxygen molecules, which make up the predominant majority of the remaining 21% of the atmosphere, are constructed similar to nitrogen molecules and thus are comprised mostly of space within their volume. See Figure 22.

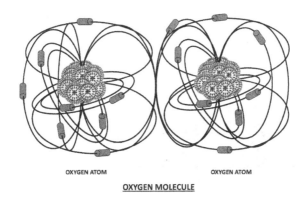

Figure 22

Illustration of an oxygen molecule.

Given that together Nitrogen (approx. 79%) and Oxygen (approx. 21%) molecules comprise 99% of Earth's atmosphere, and both molecules are comprised mostly of empty space, then most of Earth's atmosphere at the sub atomic level is empty space regardless of the fact that an atmosphere exists.

Regarding the results of the Michelson and Morley experiment, the reason the scientific community did not observe a change in the speed of light between the two light beams positioned at ninety-degree angles from each other is due to the fact that as the earth passes through space, the atoms of the earth pass through Sir Isaac Newton's luminous aether. See Figure 23. The earth is comprised of atoms all of which the vast majority of the volume of each atom is comprised of no solid material; the mass of the atom is consolidated in the nucleus of each atom. The concept that an atom represents a solid entity is a human contrived notion due to the limitation of our senses and inability to directly visualize the structure of an atom. The earth does not plow through the luminous aether creating a large wake at the surface of the planet, the earth passes through the luminous aether similar to a screen door, constructed with numerous layers of screening, passing through the air as it closes. As a solid door closes, one can feel a breeze of air brush on the skin as a volume of air is pushed out of the path of the door. As a screen door shuts, there may be a faint whisk of air due to the frame passing through the atmosphere, though generally, there is no significant disturbance detectable as such a door closes.

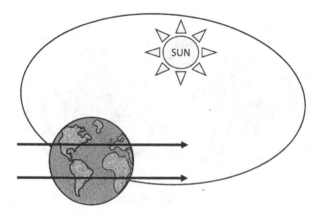

Figure 23

The luminous aether passes through the earth as the planet orbits the sun.

There is so much distance present between the nucleus of an atom and the electrons comprising the electron cloud of an atom that the vast majority of Sir Isaac Newton's luminous aether passes unimpeded through the spherical volume of an atom. See Figure 24.

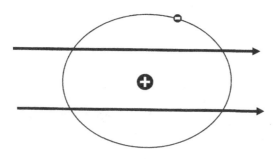

Figure 24

Majority of the luminous aether passes through the volumetric space of an atom without making contact with the nucleus or electrons of an atom.

For Michelson and Morley to have to accurately detected the presence of a wake as the earth passes through the luminous aether, they would have had to have measured the disturbance of the luminous aether on the surface of a proton or the surface of a neutron comprising the nucleus of an atom. See Figure 25. To Michelson and Morley's credit, over 130 years after their experiment, such a measurement remains beyond the capability of current science technology.

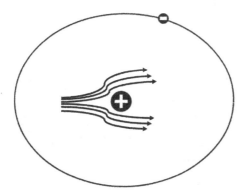

Figure 25

Measuring the wake on the surface of a proton.

An analogy to the perspective of this problem would be to consider a submerged submarine transiting through the ocean. The current perspective of the earth orbiting the sun and creating a wake, would be likened to a submarine creating a wake at the bow of the ship as the submarine moved forward. See Figure 26.

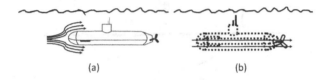

Figure 26

(a) A solid submarine maneuvering through water,
(b) Wire frame submarine moving through water.

Alternatively, if the submarine was constructed of only a thin wire frame, as the submarine moved through water, very little wake would be created due to the majority of the water passing through the body of the vessel. The wake that would be generated by the wire frame submarine would have to be measured at the points were the wire comprising the frame made contact with the water as the wire frame submarine moved forward. See Figure 27.

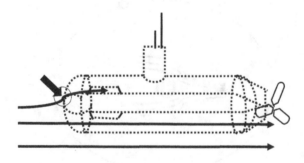

Figure 27

Measure wake at the point of wire frame makes contact with water.

The above example is NOT meant to suggest the earth, which the human senses detect, is likened to a wire frame structure. It is well understood, the earth is a solid and light does not pass through the mass of the earth that we know of by today's quantifiable standards and by what our senses detect, which registers in our conscious brain. Light is absorbed by the molecules comprising at the surface of the planet, or just under the surface of the planet. The above example is brought to the reader's attention to illustrate that the majority of the aether passing through the atmosphere is most likely traveling through the atoms comprising the atmosphere and probably through certain surface layers of the planet, if not the entire planet.

Thus, the luminous aether which acts as a medium to transport light is not appreciably affected by the passage of the earth through the aether and therefore no wake is created on the surface of the planet. If no wake is created on the surface of the planet as the earth travels along its orbital path around the sun, then the speed of light per Michelson and Morley's experimental protocol, would not be different in any direction that two test beams would be measured simultaneously. Additionally, the energy of light which originates from outside the boundaries of the planet, and strikes the planet, passes into the initial layers of matter comprising the planet, utilizing the aether, but such energy is eventually absorbed by the atoms comprising the inner matter of the planet.

Contemporary study of physics teaches that waves of the electromagnetic spectrum are reflected off the surface of molecules. Thus, white light striking a molecule and the molecule appearing to have a distinct color other than white, is due to the atoms comprising the molecules reflecting the energy, by the energy encountering the electron cloud of an atom and the energy bouncing off this electron cloud. See Figure 28.

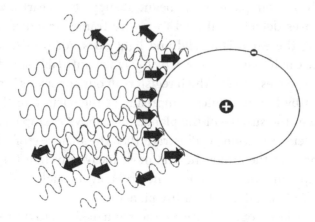

Figure 28

Contemporary view is that waves of the electromagnetic
spectrum bounce off an atom's electron cloud.

But, what if the contemporary view of the electron cloud of an atom
is incorrect, that is, light is not 'reflected' off the surface of atoms? What
if the energy that strikes an atom vibrates the nucleus of the atom, and
the vibration of the nucleus creates waves in the luminous aether which
permeates all atoms, causing what appears to be a reflection of energy.
See Figure 29.

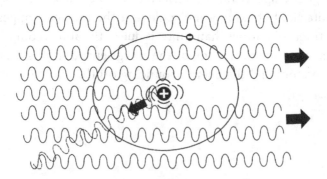

Figure 29

Energy encountering an atom strikes the nucleus causing the nucleus
to vibrate and creates a resultant wave in the luminous aether.

An analogy would be a signal buoy with a bell, floating just off the shore line, marking a channel waterway in foggy weather. If the ocean waters were still, the buoy would not rock, and no sound would be emitted by the bell. If a wave were to race across the surface of the water and strike the buoy, the result of the rocking motion of buoy would cause the bell mounted on the buoy to ring out. The more energy in the wave striking the signal buoy, the more vigorous the bell would ring and possibly the pitch of the sound emitted by the bell would change.

The classical doctrine underpinning modern physics argues 'against' the existence of a luminous aether comprised of quadsitrons, which this is supported by the fact to the common observer objects with mass look, feel, and behave as solids. The argument would be that if atoms were hollow, suspended in a three-dimensional aether, then no object should feel, look, or behave as a solid. If atoms are in essence hollow, and the electrons orbiting the nucleus of an atom do not create a solid atomic structure, then how, for example, does being struck by a baseball sailing through the air create impact, pain, and injury to the unfortunate individual hit by the baseball?

A possible analogy to support the presence of an aether is as follows: If an individual, willing to study 3-D physics is submerged in water with a breathing apparatus, such that the person is surrounded on all six sides by water, this is likened to a luminous aether which surrounds everyone all of the time. The human body is comprised of 60% water, most of which is trapped inside the cells and blood vessels comprising the tissues of the human body.[49] Thus, in effect, when the human body is completely submersed in water, this is analogous to being surrounded by a luminous aether, but also having the luminous aether inside the structure. The difference is that the exterior layer of skin of the human body represents a barrier, which defines the difference between the water inside the body and the water outside the body; therefore, there is no free flow of water from the water outside the body through the human body.

Now, to mimic an atom, one could construct a tiny ball made of a small solid center connected by wire to an exterior shell comprised of a thin wire exterior, such that water could easily flow through the structure of the wire sphere. See Figure 30. If one took this permeable wire ball and magnetized the wire, such that the ball would attach to other balls of similar construct, then such a ball could be modeled into larger structures. If one took thousands of such balls and stuck them together, such that they adhered together due to attractive magnetic forces, water would still freely permeate through the entire structure of wire balls, but the structure would exhibit mass.

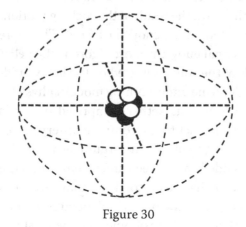

Figure 30

Wire ball with a central core.

Note, molecules are made of atoms. Molecular structures are made of hundreds, thousands, or in some cases billions of molecules. The atoms comprising molecules are bonded together by electrons flowing between atoms creating attractive forces between atoms. Molecules stick together due to bonding forces generated by the atoms comprising the molecules. A 65 kg human body is estimated to be comprised of 6.2 x 10^{27} atoms.[50] See Figure 31.

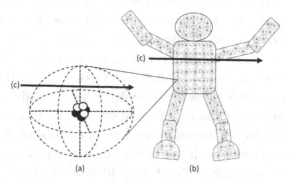

Figure 31

Atoms are comprised mostly of empty space, not occupied by subatomic structures. (a) wire model of an atom, (b)

human body comprised of atoms, (c) luminous aether passes
freely through an atom as well as the human body.

If the structure, comprised of thousands of small permeable wire balls held together by magnetic forces, struck a person submersed in water, the person would sense from the impact, that the colliding structure has mass. If the colliding structure managed to achieve a significant speed of travel through the water and if the structure was of sufficient mass, the structure comprised of thousands of small permeable wire balls might collide with the person submersed in water and cause impact, injure, or even knock the person out of the way. Thus, despite atoms being hollow, like the wire balls described above, the attractive forces generated by the electrons orbiting the nucleus of the atoms, creates the illusion of an atom being a solid structure. The description, the sense, the feel, and the behavior of any object being solid is a confabulation of the human brain due to the fact that human senses cannot directly detect the presence of the luminous aether because the elemental particles, and the behavior of such particles is so minute, we fail to be able to detect, and thus appreciate their existence.

SECTION III

REDEFINING ATOMIC PHYSICS

CHAPTER 12

QUANTA OF ENERGY

THE TERM 'QUANTA' OF energy, invites one to step into a murky quagmire of scientific knowledge on energy behavior in the universe. The observation and measurement of quantities at the quanta level are so small, that precise measurements of energy and mass are sometimes difficult to gather accurately. In addition, the actual definition of what a quanta is and what a photon is, stretches the imagination. The term 'quanta', marshalled in the age of 'quantum mechanics'. What rapidly became the dominant age of physics, replacing the older Newtonian physics, was based on the philosophy that energy in the universe could be reduced down to individual discrete packets, referred to as quanta, which could be reduced no further in energy content. How much energy is contained in a single quanta, remains elusive.

A 'photon' is an expression regarding energy, which is often interchanged with the quanta of energy. A photon represents the energy in one cycle of electromagnetic wave energy at a particular frequency.[51] Electromagnetic energy, by convention, travels at the speed of light, regardless of the frequency with which the wave is traveling at through space. Therefore, a photon travels at the speed of light, and has no mass.

Since energy is represented as numerous frequencies across the electromagnetic spectrum, differing photons may have differing amounts of energy. Planck's equation is $E = h\nu$, where E is energy in joules, h is Planck's constant and ν represents frequency of the electromagnetic wave in the equation. Planck's constant is $6.62607015 * 10^{-34}$ J-s.[52] Note Planck's constant is in joules per second, thus not in joules per packet of energy. Planck's equation demonstrates that a photon's energy is variable.

Thus, the term photon is not a candidate example for the definition of the most nonreducible unit of energy in the universe.

One can inquire, what is the resting energy of an electron. Now electrons are perpetually moving at the speed of light and never at rest, but the resting mass of an electron can be show to be 9.11 * 10 $^{-31}$ kg.[38] Given Albert Einstein provided the equation E = MC^2, knowing the mass of an object, allows one to calculate the energy content given the speed of light. An electron is shown to have an energy content of 0.501 MeV.[38] The perspective one must consider is, does the value of energy 0.501 MeV refer to the energy equivalent as related to the contents of the mass of the electron, or does the value of energy reflect the amount of energy carried solely by the mass of the electron, as suggested by the use of Einstein's E=MC^2 equation. An example might be, a basket woven from straw would have one energy signature if the straw basket were burned as a single entity. The straw basket would have quite a different energy signature if, both the straw basket and the contents of the basket, were burned in unison.

There is obviously a disassociation between an electron's energy and a photon's energy. An electron is thought to represent a form of energy, but exist with a measurable mass. A photon is thought to represent a packet of energy in the electromagnetic spectrum, but have not measurable mass.

The process of beta decay describes the transformation of a neutron to a proton. The study of the Beta decay of a free neutron demonstrates that the result of such a decay process is a proton, an electron, a neutrino and a quantity of energy. The neutrino independently represents a measurable mass and energy. The mass of three neutrinos is 0.32 ± 0.081 eV/c^2.[53] The known kinetic energy of a neutrino as a result of the Beta decay of a free neutron is 0.78 MeV.[54] See Figure 32. A naturally occurring quanta of energy might be considered to be the amount of energy associated with one neutrino, which would be calculated to be 0.1066 eV/c^2 (Eq (13)) or upon conversion 0.78 MeV. Lower amounts of energy can be measured, due to the context of the definition of what energy actually represents and the measured use of energy.

Figure 32

Concept illustration of a quanta of energy.

Detailed means to derive the energy of one quanta in joules will be discussed later in this text. This derivation will show that one quanta of free energy equals $1.7 * 10^{-20}$ joules. This value, the 'quanta', is the primary unit of free energy in the universe.

Energy represents the force that drives all action in the universe. Energy is either free as represented in a state of movement from one location to another location utilizing the density of quadsitrons as the transfer medium, or energy is captured in a loop, or energy is bound by a configuration of two or more quadsitrons. Free energy transits the universe as a single quanta, or transits the universe as a bundle of quanta of varying densities, as seen in the electromagnetic spectrum or magnetism or gravity. Captured energy is found in protons, neutrons, electrons, orbits of atoms, construct of molecules, magnetic fields. Captured energy also is trapped inside the construct of a quadsitron. Where a quanta of free energy in the universe is equal to $1.7 * 10^{-20}$ joules, the amount of energy trapped inside the structure of a single quadsitron and endlessly circulating inside a single quadsitron is $0.854494198 * 10^{-20}$ joules.

CHAPTER 13

THREE STATES OF ENERGY, PLUS ONE FALSE STATE

ENERGY IS A FORCE which is constantly in motion. Energy is always moving at the speed of light from one location in the universe to another location in the universe. Energy can be either free to move about the universe or energy can be trapped. See Figure 33. When energy is trapped, energy continues to move at the speed of light, but travels in a loop in the same location. Energy may be trapped in the orbits of electrons. Energy may be trapped in chemical bonds shared between atoms. Trapped energy flowing in loops may in some instances be appreciated by human senses such as a magnetic field surrounding a bar magnet, when two bar magnets are brought in proximity to one another and the magnetic fields collide.

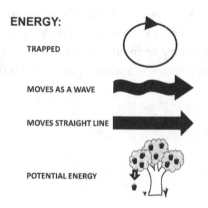

Figure 33

Energy is free to flow or is trapped

Free energy flows about the universe as a wave as seen with the electromagnetic spectrum or energy can travel straight line with little or no detectable wave pattern. When energy flows as a wave it uses the quadsitrons as a transfer medium. Energy outside the bandwidth of visible light, is not visible by the human eye. Energy trapped in a magnetic field has no detectable wave pattern. Free energy flowing with no detectable wave pattern can reach a critical density, which can make such a flow of energy capable of diverting quadsitrons parallel to its path of the energy's travel, or even move quadsitrons out of the path of the energy's travel.

There is also the subject of potential energy. Traditionally, potential energy has been considered a form of stored energy. That is, you relocate an object to a higher elevation distancing such an object further from the center of gravity of the planet earth and it has been understood that the object will somehow gain energy. An example would be that a helium balloon attached to a quarter pound weight, the helium balloon being of sufficient size to lift the quarter pound weight off the ground and up into the sky above. If the helium balloon possessed enough lift to raise the quarter pound weight off the ground and up into the air, for every increased measure of distance the quarter pound weight would rise above the starting point, most observers would believe that the quarter pound weight would gain a proportional amount of potential energy related to the height above the starting point at any given time.

If the balloon lifted the quarter pound weight up to a substantial height in the atmosphere and the balloon burst, releasing the helium and subsequently releasing the quarter pound weight, the observer would expect to see the quarter pound weight fall back to the earth. The observer would expect that the quarter pound weight would expend the potential energy it had gained by reaching the height above the starting point, falling to the earth, but in the process the quarter pound weight would gain speed at a rate of 9.80665 meters per second squared.[55] The observer would expect the quarter pound weight to strike the earth with a speed proportional to the height above the starting point the quarter pound weight had achieved on its flight provided by the now ruptured helium balloon. Such potential energy has been described as the result of an as of yet, unproven mystical force referred to as gravity.

What if the basis of the current doctrine regarding gravity were not true. What if gravity were not some mystical force that we simply have labelled with a name to satisfy our human senses, which tell us that when a balloon carrying a quarter pound weight up into the sky ruptures, the quarter pound weight falls out of the sky, and lands on the ground. What if the premise of what gravity actually represents is woefully incorrect. What if gravity were not coming from the core of the planet and pulling objects toward the core; what if gravity represented energy dense enough to stream quadsitrons out of the heavens above, toward the center of the planet and in the process created a flow, like a river, which pushed objects toward the center of the planet. Potential energy may not be some mysterious unexplained force gained by an object proportional to the object's height above a starting point, potential energy may simply be related to how an object flows in the river of quadsitrons as the three-dimensional river of quadsitrons flows from outer space down into the core of the planet.

Often a battery is considered a reservoir of energy, often referred to as potential energy. A battery is frequently considered a form of energy that is considered stationary. Instead, inside a battery, energy continues to flow, in loops, which is released when an exterior element of conductivity is attached to the battery and provides a different course of flow for the energy harbored in the battery. What is generally termed as 'potential energy' may be in fact a false state of energy, again a descriptive term to satisfy the signals the sensors of the human body are communicating to our brain regarding the environment which resides around the human form.

CHAPTER 14

NEUTRINO, TRAPPING ENERGY

ENERGY EXISTS IN ONE of two states. Energy is either trapped or energy is free. Energy which is trapped, remains in constant motion, traveling at the speed of light, but is moving within a defined boundary. Free energy travels from one location to another in space either as a wave or as a straight line. When energy moves as a wave from one location to another in space, energy is in the form of a sinus wave comprised of one or more frequencies represented in the Electromagnetic Spectrum.

The term neutrino or anti-neutrino are referrals to the presence of a primary particle in the universe. The term neutrino is used, due to the fact that such an elemental particle is observed to have no charge state. Wolfgang Pauli postulated the existence of a neutrino to explain how beta decay could conserve energy, momentum and spin.[56] There are at least three types of neutrinos reported to include electron neutrino, muon neutrino and tau neutrino. A neutrino is quite a bit smaller than other known essential particles.[57] Neutrinos are thought to pass through normal matter without being impeded or detected.[58,59] It is thought that neutrinos could be a source of dark matter.[60]

The term neutrino may not fit quite right for the following discussion and the remainder of the text, given how current physics describes this entity, but there is enough information in existence to suggest a possible loose correlation between the use of the term neutrino in this text and academic physic's use of the term.

In this work, a neutrino is theorized to be composed of two quadsitrons as presented in Figure 34. The neutrino in this figure is comprised of one left-handed quadsitrons and one right-handed quadsitrons. The quadsitron on the left side projects a negative pole toward the left sided end of the neutrino, with its positive, neutral and absolute zero poles pointed toward the inside of the neutrino. The second quadsitron on the right projects its positive pole at the opposite end of the neutrino, with its negative, neutral and absolute zero poles pointed toward the inside of the neutrino. Inside the neutrino, the positive pole of the left neutrino is attracted to the negative pole of the right neutrino, while the neutral pole and absolute zero poles of each quadsitron attach to each other. Cradled within two quadsitrons, which comprise a neutrino, is a quantity of energy. The amount of energy trapped inside the core of the neutrino is calculated to be equal to 0.1066 eV/c² (Eq (13)).

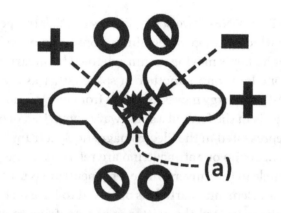

Figure 34

Illustration of a neutrino comprised of two quadsitrons, one with positive pole pointed outward, the opposite quadsitron with the negative pole pointed outward, then internally the negative pole and positive pole face each other, while the neutral poles are in proximity to the absolute zero pole of the opposing quadsitron.
(a) Quanta of energy captured between two quadsitrons.

The neutrino presented in Figure 35 demonstrates the field lines, which surround the merging of a left-handed quadsitron and a right-handed quadsitron.

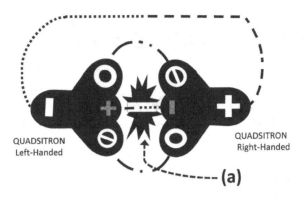

Figure 35

Illustration of the field lines surrounding a neutrino comprised of two quadsitrons, one with positive pole pointed outward, the opposite quadsitron with the negative pole pointed outward, then internally the negative pole and positive pole face each other, while the neutral poles are in proximity to the absolute zero pole of the opposing quadsitron. (a) Quanta of energy captured between two quadsitrons.

A neutrino in the context of this treatise, consists of two quadsitrons and a quanta of energy. The quanta of energy refers to $1.7 * 10^{-20}$ joules of free energy, which has been trapped by the union of two quadsitrons. This $1.7 * 10^{-20}$ joules of free energy combines with, circulates and strengthens the field lines already present in the two quadistrons. Each quadsitron has an individual energy of $0.854494198 * 10^{-20}$ joules, which is nonreducible and cannot leave the quadsitron. Thus, a neutrino has a combined energy of $3.4 * 10^{-20}$ joules, which is the sum of $1.7 * 10^{-20}$ joules of free energy and $0.854494198 * 10^{-20}$ joules held bound in each of the two quadsitrons.

Neutrinos represent building blocks of protons and neutrons. The colossal massiveness and density of the proton and neutron in comparison to the quadsitron and neutrinos allows protons and neutrons to freely pass through the three-dimensional aether comprised of free quadsitrons and neutrinos, which fill the three-dimensional space of the universe.

Neutrinos come in differing forms depending upon how the poles of the quadsitrons are aligned when the neutrino is created. Neutrinos are most likely formed in the highly charged compressive strength of the vortex of a galaxy. The above description is that of a polar neutrino. A nonpolar neutrino would be representative of Dark Matter.

CHAPTER 15

DARK MATTER

DARK MATTER IS PROBABLY the most alluring, mysterious material in the universe. It is thought that Dark Matter comprises 27% of the volume of the known universe.[61] Dark Matter is considered to be material, but relatively undetectable by standard methods of measurement. Dark Matter may be undetectable due to the conventional positive and negative poles of this uniquely structured neutrino being concealed within the structure. The existence of a Dark Matter neutrino is the result of a particular combination of a left-handed quadsitron and a right-handed quadsitron.

Dark Matter is comprised of taking two quadsitrons, one right-handed and one left-handed, and merging them together into a neutrino, which bears an absolute zero pole to one side and a neutral pole to the opposing side. Between the two quadsitrons is trapped a quanta of energy. The positive and negative poles of the quadsitrons are merged together on the interior of the neutrino. See Figure 36. The close proximity of the positive and negative poles may create a very stable dark matter neutrino.

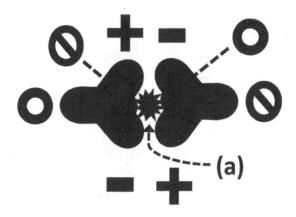

Figure 36

Dark Matter with one right-handed quadsitron and one left-handed quadsitron merged together into a special neutrino which bears an absolute zero pole to one side and a neutral pole to the opposing side. (a) represents a quanta of energy.

Clay was used to model a left-handed quadsitron and a right-handed quadsitron. Presented in Figure 37, is the colored (gray scaled for purposes of this book) clay model of a left-handed quadsitron merged with the model of a right-handed quadsitron. The 3D clay models in Figure 37 demonstrate the appearance of Dark Matter with the positive poles and negative poles of the quadsitrons concealed inside the structure. The absolute zero pole of the right-handed quadsitron is pointed outward and to the left, while the neutral pole is concealed; the absolute zero pole of the left-handed quadsitron is concealed, while the neutral pole is pointed outward and to the right of the figure. A quanta of energy is captured within the interior of the Dark Matter neutrino.

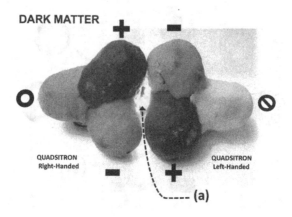

Figure 37

Clay models of the merging of a left-handed and a right-handed quadsitron, with absolute zero pole pointed to the left and neutral pole pointed to the right, to construct a neutrino to represent Dark Matter. (a) represents a quanta of energy.

Dark Matter, like the polar neutrino discussed in Chapter Fourteen with one positive pole exposed and one negative pole exposed, trap a quanta of energy. But energy is thought to never stop moving at the speed of light. Quadsitrons have an inherent energy which creates weak field lines that surround the quadsitron structure. See Figure 38.

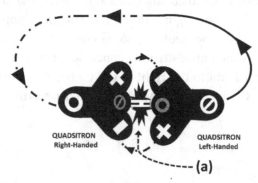

Figure 38

Dark Matter, the exterior field lines surrounding the two quadsitrons are strengthened by the quanta of energy trapped between the two quadsitrons. (a) represents a quanta of energy.

When a neutrino is formed, whether a polar neutrino or a Dark Matter neutrino, the quanta of energy trapped by the merger of the two quadsitrons circulates through both quadsitrons and strengthens the field lines which surround the two quadsitrons. A polar neutrino becomes more energized and thus more polar. A Dark Matter neutrino also becomes more energized, but remains detectably silent by conventional means of discovery.

CHAPTER 16

CONSTRUCT OF A PROTON

PROTONS COMPRISE THE NUCLEUS of every atom. Generally, there are equal numbers of protons and electrons for any given atom. A proton is theorized to be a sphere comprised of a volume of b = 1.75926*10^{10} quadsitrons Eq (113). The radius of the proton is calculated to be comprised of 1300 quadsitrons Eq (137). A proton is theorized to be comprised of a sphere of neutrinos. The majority of the neutrinos are polarized. At the center of a proton is Dark Matter, due to the neutral effect of the neutrino required at the core of the subatomic particle. See Figure 39.

The entity of Dark Matter is theorized to be comprised of one left-handed quadsitron and one right-handed quadsitron oriented with one absolute zero pole pointed outward and one neutral pole pointed outward, with a quanta of energy trapped between the two quadsitrons. The positive pole of each quadsitron is attracted to the negative pole of the opposing quadsitron.

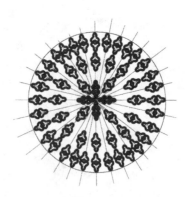

Figure 39

Concept drawing of a proton.

A proton is a sphere comprised of neutrinos. The radius of a proton is comprised of 650 neutrinos Eq (141), the majority of the neutrinos comprising a proton are polarized. Within a proton, the polarized neutrinos position their negative pole directed toward the center of the proton and their positive pole pointed toward the exterior of the proton.

The quanta of energy trapped by each neutrino creates intensified exterior field lines. The energy of the exterior field lines results in a circulation of energy throughout the entire sphere. The point where the structure becomes unstable is at a radius of 650 neutrinos distant from the center of the proton. The entire exterior surface of the proton is comprised of the positive pole of the surface neutrinos. The attractive forces between polar neutrinos become borderline at a sphere size of 650 neutrinos from the center of the proton, which facilitates the construct of the structure known as a neutron.

CHAPTER 17

CONSTRUCT OF A NEUTRON

NEUTRONS COMPRISE THE NUCLEUS of all atoms, except for hydrogen atom which has no neutron, only a lone proton. The remaining elements (2-118) of the Periodic Table are constructed with generally an equal number of protons and electrons. The number of protons is the means of distinguishing the atom of one element, from an atom of a differing element. It is thought, generally, there are an equal number of protons and neutrons present in the nucleus of an atom, but this concept is often variable in nature to order to insure stability of an atom's structue.[62] Isotopes of an element will vary from this format containing more or less the number of neutrons in comparison to the number of protons in the nucleus of the atom. Often a naturally occurring forms of an element are more stable if constructed with differing number of protons and neutrons occupying the nucleus.

A neutron is comprised of a volume of $1.76168 * 10^{10}$ quadsitrons Equation (108). A neutron is theorized to be a sphere with a radius of 1301 quadsitrons Equation (143) as illustrated in Fig. 40. The difference between the spherical composition of a proton and the spherical composition of a neutron is a single spherical layer of quadsitrons.

Figure 40

Concept illustration of a neutron

The difference of the radius of a neutron versus the radius of a proton is 5.11104551 * 10^{-19} m Eq (131). This difference in the radius of a neutron and the radius of a proton is smaller than the diameter of a quadsitron 8.557941465 * 10^{-19} m Eq (79), representing sixty percent of the diameter of a quadsitron. The reason for this discrepancy is that the exterior layer of quadsitrons, which attach to the exterior of the proton are a shell of single layer of quadsitrons, which cradle the exposed positive pole of the polar neutrinos located on the surface of the proton. The neutrinos on the surface of the proton project their positive pole outward from the center of the proton. The layer of quadsitrons that cover the layer of neutrinos to transform a proton into a neutron project their neutral pole outward, away from the center of the neutron. Due to the angle of 120 degrees that separates each of the four poles of a quadsitron, this allows a free quadsitron to sit cradled on top of the bound quadsitron comprising the neutrino on the surface of the proton, with a forty percent by volume overlap of the two quadsitrons as shown in Fig. 41.

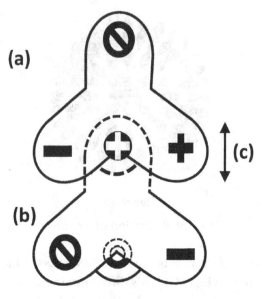

Figure 41

Concept drawing of a quadsitron overlapping and resting on top of a quadsitron on the surface of a proton to create a neutron. (a) Quadsitron acting as the surface layer of the neutron with the neutral pole pointed outward. (b) Quadsitron located on the surface of a proton as part of the neutrino on the surface of the proton with the positive pole pointed outward. (c) Forty percent overlap of the surface quadsitron of the neutron and the surface quadsitron of the proton.

The process of a neutron transforming into (1) a proton, (2) an electron, and (3) a neutrino is referred to a Beta Decay of a neutron.

CHAPTER 18

EXPLAINING BETA DECAY

THERE EXISTS A TENUOUS relationship between a free quadsitron attaching to the surface of the proton to generate a neutron with the quadsitron bound to a second quadsitron comprising a neutrino at the surface of a proton. The single layer of free quadsitrons which create the neutron have their positive and negative poles pointed toward the quadsitrons positioned on the surface of the proton. The negative, positive, absolute zero poles of the free quadsitron overlap the positive pole of the bound quadsitron as the free quadsitron cradles the bound quadsitron. The negative pole of the free quadsitron is attracted to the positive pole of the bound quadsitron. The positive pole of the free quadsitron is repulsed by the positive pole of the bound quadsitron, but the positive pole of the free quadsitron is also simultaneously attracted to the negative pole of the bound quadsitron, which creates the tenuous attractive relationship between the free quadsitron and the bound quadsitron. The energy trapped by the conversion process of a proton transforming into a neutron provides the energy to bind the single layer of free quadsitrons to the bound quadsitrons on the surface of the proton by overcoming the repulsive force of the positive pole of free quadsitron and the positive pole of the bound quadsitron. The unstable relationship of the positive pole of the free quadsitron and the positive pole of the bound quadsitron on the surface of the proton in the proper set of circumstances, facilitates the shedding of the superficial layer of quadsitrons to transform a neutron to a proton.

The concept of Beta decay is illustrated in Figure 42 with the most exterior layer of quadsitrons separating from the body of the underlying proton.

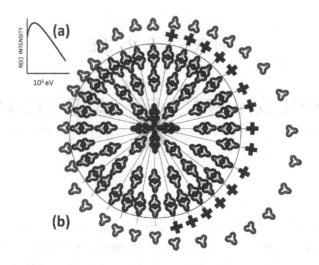

Figure 42

Neutron in Beta decay to form a proton. (a) Beta-ray spectrum
described in Beta decay of radium E. (b) Exterior layer of
quadsitrons separating from a free neutron to create a proton.

A neutron shedding a single layer of quadsitrons to produce a proton
exhibits an energy signature which initially rises to a maximum and
then declines. The conversion of a neutron to a proton with the neutron
shedding a single layer of quadsitrons to produce a proton is likened to
the peeling an orange or other similar fruit. If the removal of the surface
of a free neutron is initiated at one location and spreads uniformly across
the surface of the spherical shaped neutron, the energy released as the
surface layer of the sphere peels away, spreading in all directions, will rise
to a maximum as the involved surface area reaches the greatest surface
area to be involved, then the release of energy will decline as the involved
area of shedding of the surface layer diminishes and reaches its finale.
The curve of a Beta-ray spectrum such as described by Neary in The
Beta-ray spectrum of radium E, demonstrates the spectrum of energy
emitted by a radioactive substance (radium E).[42] Radium E was a term
previously used to describe Bismuth, atomic weight 210, which emits
Beta radiation. The energy emitted by a mass of neutrons decaying at
differing times to protons, such as a quantity of Bismuth-210, will release

energy to correspond to the energy signature which initially rises to a maximum and then declines.

Thus, a neutron transforms into a proton by shedding a single layer of quadsitrons from the surface of its sphere. This single layer of quadsitrons is held tenuously to the surface of polar neutrinos by bonding created by a flow of energy between the quasi-free quadsitrons on the surface and bound quadsitrons comprising the proton underneath. With the shedding of a layer of quadsitrons and dispersal of enough energy to generate an electron and a neutrino, the physical products of Beta decay becomes a proton, an electron, and a neutrino.

CHAPTER 19

HOW DO ALL OF THOSE ELECTRONS NOT COLLIDE WITHIN ATOMIC STRUCTURES?

WE HAVE BEEN LED to believe that an electron is some form of solid structure. Thus, when an electron collides with something it strikes with force due to the electron's speed and mass. We have also been led to believe that atoms, and therefore molecules which are constructed from atoms, are solid due to the mass and speed of electrons as the electrons orbit the center of an atom at light speed. It has been theorized that the electron traveling at the speed of light, orbits the center of an atom so fast, that it is as if the electron is in all places in the orbit at all times, thus creating a presence of solid exterior about the center of the atom.

The description of the electron moving so fast in its orbit around the center of an atom creating the presence of solid mass about the center of an atom runs into trouble when one starts to study the behavior of electrons and the construct of the electron cloud about atoms comprised of numerous protons and neutrons at the center of the atom.

Electrons are thought to be the instrument that captures and stores energy for an atom. As an atom absorbs free energy one or more electrons occupying the outer orbits of the atom distance themselves further from the center of the atom.[15] This is referred to as a valance electron. Such valance electrons, occupying the outer most shell of an atom, also participate in chemical bonding between atoms and dictate the chemical behavior of an element.

Heightened energy levels of an atom are measured in individual quanta absorbed by the atom. Individual quanta absorbed by an atom translates into an integer distance the electron(s) in the outer shell orbits are from the center of the atom. When an atom releases a quanta of energy, an electron in an outer orbit of the atom moves closer to the center of the atom by an integer distance equal to the energy of the released quanta of energy. See Figure 43.

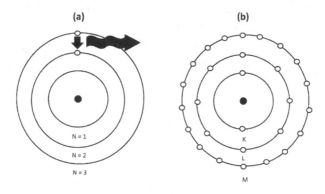

Figure 43

(a) As atoms release a quanta of energy, an orbital electron moves closer to the center of the atom by an integer equal to the quanta released. (b) Atoms can be illustrated by dot diagrams to represent the presence of electrons in differing orbital shells about the center of the atom.

Niels Bohr first described the atom as having a nucleus at the center with electron orbitals circulating around the nucleus. Since Bohr's original description, the suggested construct of the atom has extensively evolved. Atoms have been represented by dot diagrams to indicate the presence of electrons in differing orbital shells about the center of the atom. The dot diagram is representative of the original Bohr model of an atom, when electron orbitals were thought to be spherical in their construct similar to planets orbiting a star. The early rendition of the atom suggested there existed a K orbital, an L orbital and an M orbital. Often a number of electrons were assigned to each orbital. The K orbital, the orbital closest to the nucleus, was thought to be occupied by two electrons. The L orbital was thought to be occupied by eight electrons. The M orbital was thought to have a maximum of eighteen electrons, which could occupy the spherical orbital.

The current atomic model of an atom describes numerous orbitals surrounding the nucleus of an atom to account for all of the electrons comprising the atom and is often referred to as the Electron Shell Model of the atom.[16] The orbitals comprising the Electron Shell Model of the atom are referenced quantum numbers with suborbitals reported as orbital s, orbital p, orbital d, and orbital f.[63] Each of the orbitals s, p, d, f are shaped differently to account for the electron dispersal pattern which is thought to make up the orbitals.[64]

The 's' orbital appears like a sphere. Presented in Figure 44 is an illustration of the 's' orbital. Two electrons are capable of occupying an 's' orbital.

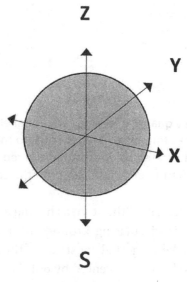

Figure 44

s-orbitals.

The 'p' orbital appears like a figure eight or dumbbell in shape, which may occupy the x, y, and/or z axis. Presented in Figure 45 is an illustration of the 'p' orbitals. Up to six electrons may occupy the p-orbitals projecting in the x, y, and z axis.

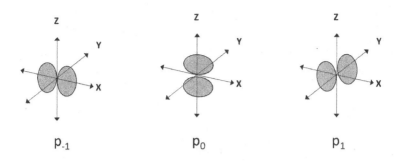

Figure 45

p-orbitals.

Of the 'p-orbitals', the p_{-1} set are two orbitals oriented hypothetically along the x axis. The p_0 set would be oriented along the z axis. The p_1 set would be oriented along the y axis. Each set of orbitals could accommodate two electrons, for a maximum of six electrons allocated to the p-orbitals.

The 'd-orbital' has several shapes but generally appear as pear fruit, occupying the x, y, and/or z axis. Presented in Figure 46 is an illustration of the d-orbitals. Up to ten electrons may occupy the d-orbitals.

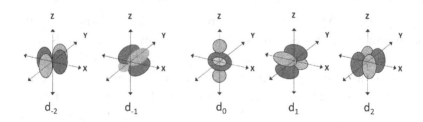

Figure 46

d-orbitals.

The 'f-orbitals' have several shapes, occupying the x, y, and/or z axis. Presented in Figure 47 is an illustration of the f-orbitals. There are seven orbitals comprising the f-orbitals, two electrons per orbital, therefore up to fourteen electrons may occupy the f-orbitals.

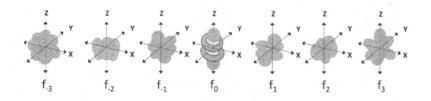

Figure 47

f-orbitals.

Presented in Figure 48 is an illustration summarizing the recognized orbitals utilized to create the known elements. There are seven known layers or shells of orbitals. The first shell, closest to the nucleus of the atom, contains only an 's' orbital; maximum electrons capable of occupying the first layer is 2; this is designated as $1s^2$. The second shell, second closest to the nucleus of the atom, contains an 's' and a 'p' orbital; maximum number of electrons capable of occupying the second shell is 8; this is designated as $2s^2$, $2p^6$. The third shell from the nucleus of the atoms, contains an 's', 'p' and 'd' orbitals and maximum number of electrons capable of occupying the third shell is 18; this is designated as $3s^2$, $3p^6$, $3d^{10}$. The fourth shell from the nucleus of the atoms, contains 's', 'p', 'd' and 'f' orbitals, maximum number of electrons capable of occupying the fourth shell is 32; this is designated as $4s^2$, $4p^6$, $4d^{10}$, $4f^{14}$. The fifth shell theoretically contains 's', 'p', 'd', 'f' and 'g' orbitals, maximum number of electrons capable of occupying the fifth shell is 50; this is designated as $5s^2$, $5p^6$, $5d^{10}$, $5f^{14}$, $5g^{18}$. The theory of how elements are constructed using the Electron Shell Model, begins to break down as the nucleus of an atom becomes denser with the addition of protons and the number of electrons added to the outer orbitals increases with the larger atomic structures in the Periodic Table.

There is no known element which has more than 32 electrons residing in one shell.[65] Thus, the 'g-orbitals' are speculative. In the larger elements, some of the outer orbitals fill before the inner orbitals are full. The sixth shell is known to support an 's', 'p' and 'd' orbitals. The seventh shell is known to support an 's' and 'p' orbital. The construct of the largest elements is more theoretically abstract.

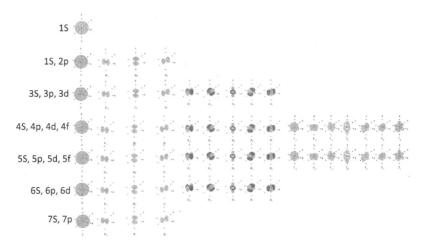

Figure 48

Summary of potential orbitals which may comprise an atom.

Copper (Cu) has an atomic number of 29. Most commonly in nature, the copper atom contains 29 protons, 34 neutrons, and 29 electrons.[66] As mentioned earlier, often in nature, atoms of an element may be found with unequal number of protons and neutrons present in the nucleus of an atom. Atomic mass of copper is 63.546 atomic mass unit (amu). Copper has two stable isotopes Cu-63 and Cu-65. Cu-63 comprises 69% of the naturally occurring copper.[67]

The copper atom has four levels of energy. See Figure 49. To accommodate the 29 electrons, the copper atom is thought to be constructed with 1s, 2s, 2p, 3s, 3p, 3d and 4s orbitals.[66] The 1s, 2s, 2p, 3s, 3p, and 3d orbitals are full. The 4s orbital is occupied by one electron. Commonly accepted notation is as follows: $1s^2$, $2s^2$, $2p^6$, $3s^2$, $3p^6$, $3d^{10}$, $4s^1$.[68] Copper is utilized as a general conductor of electricity in electric circuits. The single electron occupying the 4s orbital is allowed to participate in the flow of current (electrons), when a potential energy (battery) or active energy (wall plug connected to main power) source is connected to a copper wire and a completed circuit with appropriate resistance is established.

COPPER
CU
Atomic Number 29

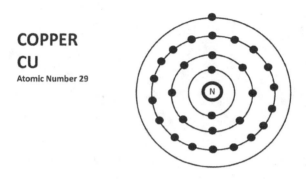

Figure 49

Orbitals comprising a copper atom using Bohr configuration.

Gold proves to be a superior conductor of electricity compared to copper. The relative scarcity and the much higher cost of gold compared to copper, makes copper the common source of constructing electric circuits. If the price of gold and copper were the same, gold would be the choice for wiring technical electric circuits.

Gold (Au) has an atomic number of 79. The most common form of gold, is gold-197.[69] This stable form of gold contains 79 protons, 118 neutrons, and 79 electrons. Gold's atomic weight is 196.96655 amu. Gold possesses six energy shells.

To accommodate the 79 electrons, the gold atom is thought to be constructed with a 1s, 2s, 2p, 3s, 3p, 3d, 4s, 4p, 4d, 4f, 5s, 5p, 5d, and 6s orbitals. See Figure 50. The 1s, 2s, 2p, 3s, 3p, 3d, 4s, 4p, 4d, 4f, 5s, and 5p orbitals are full. The 5d orbital is partially full. The 6s orbital is occupied by one electron. Commonly accepted notation for the gold orbitals is as follows: $1s^2$, $2s^2$, $2p^6$, $3s^2$, $3p^6$, $3d^{10}$, $4s^2$, $4p^6$, $4d^{10}$, $4f^{14}$, $5s^2$, $5p^6$, $6s^1$, $5d^{10}$. The single electron occupying the 6s orbital is allowed to participate in the flow of current (electrons), when a potential energy or active source is connected across a gold wire and a completed circuit with appropriate resistance is established. Theoretically, since one electron which occupies the 6s orbital in the gold atom, which is further from the nucleus of the gold atom than the electron in the 4s orbital of a copper atom, gold is a better conductor of electricity than copper. The hypothesis is that the 6s orbital allows more freedom to the mobility of the distant electron, being further away from the nucleus of the gold atom. The 6s orbital allows for electrons to hop from one gold atom to another gold atom with less resistance to current flow, than copper atoms allow in their 4s orbitals.

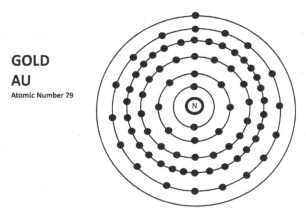

Figure 50

Orbitals comprising a gold atom utilizing the Bohr configuration.

The current theory of the construct of an atom involves layers of the s, p, d, f orbitals as described above. Illustrated in Figure 51, are the multiple overlapping layers of electron orbitals that would be required to construct a copper atom.

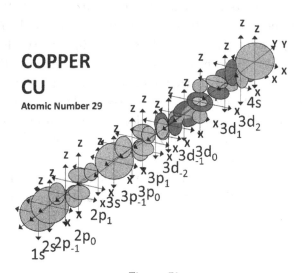

Figure 51

Layers of orbitals comprising a copper atom.

A gold atom would have a far more complex set of orbitals as compared to copper seen in Figure 51. The orbital geometry of a gold atom would need to account for 79 electrons, rather than copper's 29 electrons. The gold atom would require the addition of $4s^2$, $4p^6$, $4d^{10}$, $4f^{14}$, $5s^2$, $5p^6$, $5d^{10}$ and $6s^1$ orbitals to the orbitals already depicted in Figure 51. The complexity of the atomic orbitals becomes almost astronomical.

Intuitively, the layers of orbitals illustrated in Figure 51, conflict with the current teachings of the structure of an atom. The reason solid substances are 'solid' is thought to be because of the behavior of electrons orbiting the center of an atom. Electrons are thought to be both having the quality of being solid and to be moving at the speed of light. The fact that electrons move at the speed of light, is thought to create the physical appearance that the electron in an orbital occupies all positions in a particular orbital at the same time, thus creating the illusion of the substance of matter. If there are multiple layers of orbitals surrounding the nucleus of an atom, it would stand to reason, if the behavior of electrons is to create a simulated solid substance, electrons would crash into each other with catastrophic results.

Given the difficulty of properly explaining the orbitals of an atom without having electrons constantly crashing into each other, a traditional position that electrons are solid objects comes into question. Albert Einstein theorized that solid objects could get close to the velocity of the speed of light; but as the speed of light was approached the mass increased. Objects could not reach the speed of light due to the increase in relativistic mass of the object.[70] At the speed of light, an entity is pure energy, but also possibly theorized to be of infinite mass. It is likely then that electrons are not solid objects. It is possible electrons, instead, are a phenomenon created by quanta channeled into orbits around the nucleus of an atom, these channels organized by arrangements of quadsitrons, and these arrangements of orbitals is orchestrated by the collection protons and neutrons present in the nucleus of the atom exerting forces upon the three-dimensional field of quadsitrons surrounding the nucleus.

CHAPTER 20

EXPLAINING THE CONSTRUCT OF THE ELUSIVE ELECTRON

THE CURRENT ELECTRON SHELL Model of an atom has been derived and evolved extensively over time.[16] We have been led to believe that the electron is an object. We have been told that an atom is comprised of a nucleus made up of protons and neutrons located at the center, and electrons occupy orbitals around the nucleus. We have been told that an electron travels at the speed of light and possesses mass. We have also been told that objects traveling at the speed of light transform into pure energy and thus, have no mass. Therefore, we have in essence been taught that an electron defies the laws of physics by traveling at the speed of light, yet possess mass. Somewhere in our description of an electron or our understanding of physics, there exists a critical conflict.

Possibly, an electron is not a solid object with mass, but at any given instance in time, represents a quanta of energy and has the illusion of having a mass. To investigate this hypothesis of an electron being a solid versus an electron having the illusion of mass, one can return to the study of a neutron undergoing Beta Decay. As mentioned previously in this text, a neutron undergoing Beta Decay generates a proton, a single electron and a neutrino. In a universe where quadsitrons represent the most elemental form of matter, and Beta Decay is explained by a single layer of quadsitrons separating from the surface of a neutron, the single layer of quadsitrons, minus two quadsitrons to construct the neutrino, is representative of an electron.

Investigating the surface layer of quadsitrons covering a neutron, is simply an exercise in special mathematics. Knowing the radius of a neutron, knowing the diameter of a quadsitron, and utilizing the mathematics of geometry, the number of quadsitrons which comprise the surface layer of a neutron can be calculated.

Knowing the estimated diameter of an electron, and using the calculated size of a quadsitron as a unit of reference, the number of quadsitrons comprising an electron can be investigated to determine the shape characteristics of an electron. The quadsitrons, used in this case as a unit of mass and a unit of size (per calculated diameter and approximating the structure of a quadsitrons as a sphere) offers the instrument to dissect the known mass of an electron, the known diameter of an electron and the known quantity of mass and calculated number of quadsitrons distinguishing the electron component from a neutron when a neutron undergoes the process of Beta Decay.

Correlating the calculated size of the individual quadsitron with the calculated number of quadsitrons as a product of Beta Decay, these can be used as references to investigate the estimated size of an electron. Approximating that an electron is a sphere allows one to investigate whether there is enough matter as a product of neutron Beta Decay to construct an electron as a solid sphere, as a hollow sphere, or as a series of rings. The investigation into the geometry of an electron is presented below.

STEP #1 Solve for the Number of Quadsitrons Available to Comprise an Electron

M_e = 9.10938356 x 10^{-31} kg, where M_e represents electron weight or mass. [Reference 38]

The weight of a quadsitron is 9.5075301706 x 10^{-38} kg

M_e = g x M_q, where given the known weight of an electron and known weight of a quadsitron, 'g' is number of quadsitrons being solved for and M_q is the weight of a single quadsitron.

g = M_e / M_q

g = 9.10938356 x 10^{-31} kg / 9.5075301706 x 10^{-38} kg

g = 9.581230 x 10^6

The number of quadsitrons available by weight to comprise an electron is 9.581230×10^{6}.

STEP #2 Solve for the Number of Quadsitrons That Would Comprise Electron Volume if an Electron Were a Solid Sphere

Classical electron radius = Re = 2.81794×10^{-15} m

Evol = U * Qvol

where U is the theoretical number of quadsitrons that would be present in an electron, if an electron were a solid sphere, and if quadsitrons could be packed tightly together with no separation of space, Evol is the volume of an electron, Qvol is the volume of a quadsitron.

Volume of an electron = Evol= $4/3 * \pi * (Re)^3$

Evol= $4/3 * \pi * (2.81794 \times 10^{-15}$ m$)^3$

Evol= $93.731126080 \times 10^{-45}$ m^3

Radius of a quadsitrons = Rq

Rq = $4.2785494790 \times 10^{-19}$ m

Volume of a quadsitron = Qvol = $4/3 * \pi * (Rq)^3$

Qvol = $4/3 * \pi * (Rq = 4.278549479 \times 10^{-19}$ m$)^3$

Qvol = $3.280788888658 \times 10^{-55}$ m^3

U = Evol / Qvol

U = $93.731126080 \times 10^{-45}$ m^3 / $3.280788888658 \times 10^{-55}$ m^3

U = $28.56969139466 \times 10^{10}$ quadsitrons

The number of quadsitrons by weight calculated to comprise an electron is 9.581230×10^6. To create a solid sphere the number of quadsitrons needed would be $28.56969139466 \times 10^{10}$, therefore, an electron cannot be a solid sphere comprised of quadsitron derived from the resultant mass change of the neutron decaying to a proton. There is an insufficient number of quadsitrons generated by a neutron decaying to a proton to create a solid sphere the size of an electron.

STEP #3 Solve for the Number of Quadsitrons That Would Comprise Electron Volume if an Electron Were the Surface Shell of a Sphere

Electron radius = Re = 2.81794×10^{-15} m [Reference 71]

Surface area of an electron = Esur = U' * Qarea

where U' is the theoretical number of quadsitrons that would be present in an electron, if an electron were a hollow sphere, and if quadsitrons could be packed tightly together with no intervening space on the surface of the sphere.

Surface of an electron = Esur = $4 * \pi * (Re)^2$

Esur = $4 * \pi * (Re)^2$

Esur = $4 * \pi * (2.81794 \times 10^{-15}$ m$)^2$

Esur = $9.978685786 \times 10^{-29}$ m^2

Rq = $4.2785494790 \times 10^{-19}$ m

Area taken at center of a quadsitron = Qarea = $\pi * (Rq)^2$

Qarea = $\pi * (4.2785494790 \times 10^{-19}$ m$)^2$

Qarea = $5.75099500059 \times 10^{-37}$ m^2

U' = Esur / Qarea

U' = 9.978685786 x 10^{-29} m^2 / 5.75099500059 x 10^{-37} m^2

U' = 1.735123362996 x 10^8 quadsitrons

    The number of quadsitrons by weight calculated to comprise an electron is 9.581230 x 10^6. To create a hollow sphere the number of quadsitrons needed would be 1.735123362996 x 10^8, Therefore, an electron cannot be a hollow sphere comprised of quadsitrons derived from the resultant mass change of the neutron decaying to a proton. There is an insufficient number of quadsitrons generated by a neutron decaying to a proton to create a hollow sphere the size of an electron.

STEP #4 Solve for the Number of Quadsitrons That Would Comprise Electron Volume if an Electron Were a Series of Hollow Rings

Electron radius = Re = 2.81794 x 10^{-15} m

Quadsitron diameter = Dq = 8.557098 x 10^{-19} m

Circumference of an electron = Ecir = 2 * π * Re

    where U$^{\#}$ is the theoretical number of quadsitrons that would be present comprising an electron, if an electron were one hollow ring:

Ecir = 2 * π * (2.81794 x 10^{-15} m)

Ecir = 17.7056392011 x 10^{-15} m

Ecir = 1.77056392011 x 10^{-14} m

Dq = 8.557098 x 10^{-19} m

U$^{\#}$ = Ecir / Dq

$U^{\#} = 1.77056392011 \times 10^{-14}$ m / 8.557098×10^{-19} m

$U^{\#} = 20,690$ quadsitrons

The number of quadsitrons by weight calculated to comprise an electron is 9.581230×10^{6}. To create a single hollow ring, the number of quadsitrons needed would be 20,690. Therefore, an electron cannot be one single hollow ring.

Number of hollow rings = 9,581,230 / 20,690

Result of the above division is the number of hollow rings = 463.085 hollow rings

An electron could be comprised of 463 hollow rings.

If an electron is comprised of 463 hollow rings, then an electron is 463 * diameter of a quadsitron = 463 * 8.557098×10^{-19} m = $3.96193637 \times 10^{-16}$ m in length while having a circular radius of 2.81794×10^{-15} m. See Figure 52.

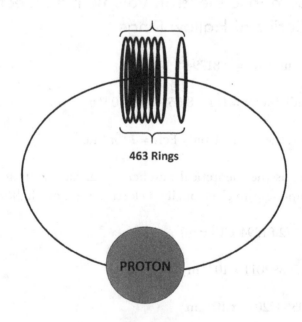

463 Rings

PROTON

Figure 52

Concept drawing of an electron consisting of 463 hollow rings.

Therefore, an electron may simply be a measure of 463 rings of quadsitrons, which represents the length of a quanta of energy comprising an electron at any instant in time. The construct of protons and neutrons at the center of an atom may create potentials in the surrounding aether that act like roadways for quanta of energy to flow in an organized orbital around the nucleus of an atom.

At any given point in time, an electron, as defined as an entity in orbit around the nucleus of an atom, may consist of 463 rings of quadsitrons and a quantum of energy. Like roads automobiles drive upon, concrete or asphalt roads exist to be potentially used by an automobile, or multiple automobiles. At any given time, a road is present and ready to service a car, but at any given time there may or may not be an automobile present on a given segment of road.

Electrons defined as a width of 463 hollow rings containing a quanta of energy, is capable of traveling at the speed of light and have a measurable mass at the same time and not conflict with the theory that at the speed of light an object has no mass. The segment of rings surrounding the quanta of energy change per increment of time as the quanta of energy orbits the nucleus of the atom. The quanta of energy associated with an electron travels at the speed of light. The mass of the electron at any given instant in time is equivalent to 9.58×10^6 quadsitrons or as calculated, a sequence of 463 rings of quadsitrons with the diameter of an electron.

The current Electron Shell Model of particle theory has evolved over time as an effort to refine the teachings of the Rutherford-Bohr Model. The hypothetical construct of an electron existing as 463 rings of quadsitrons and a quanta of energy explains how numerous electrons are able to exist as numerous overlapping electron orbitals in the Modern model of the atom with its varying electron orbitals. If an electron is, in fact, a quanta of energy with an associated set of rings of quadsitrons at any given point in time, the quanta of energy following an orbital pathway dictated by the forces created in the luminous aether by the construct of the protons and neutrons comprising the nucleus of the atom, the quanta of energy moving, but the quadsitrons staying stationary, potentially responding to the next electron passing through the same space, this concept facilitates that electrons can co-exist in the same space surrounding the nucleus of an atom.

In addition, there may be a portion of the one or more quanta of energy associated with an electron which circulates from the orbit of the electron, down a path directly to the nucleus of the atom, then back up a path to the electron as the electron orbits the center of the atom. See Figure 53. The quanta of energy associated with circulating from the electron's position in orbit to the nucleus of an atom would provide a measure of stability for the quadsitrons comprising the shape of the electron and explain how an electron, occupying an outer orbital of an atom, can absorb quanta of energy in integer quanta when an atom absorbs energy.

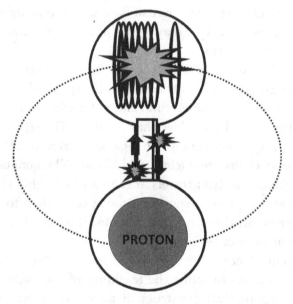

Figure 53

Quanta of energy circulates from an electron to the center of an atom.

Electron Spin

Given that there are 463 rings of quadsitrons there is nothing to say that energy in a quanta cannot make the quadsitrons spin as the quanta of energy orbits the nucleus of an atom. Energy of sufficient density can move quadsitrons, energy of sufficient density can spin quadsitrons and can spin rings of quadsitrons. Quanta of energy may spin the rings of quadsitrons as a right-handed spin or a left-handed spin. See Figure 54.

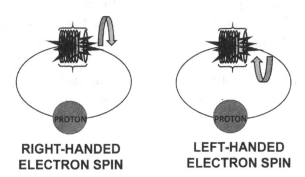

**RIGHT-HANDED
ELECTRON SPIN**

**LEFT-HANDED
ELECTRON SPIN**

Figure 54

Right-handed and Left-handed electron spin.

The spin of the electron assists in the compatibility of multiple electrons in orbit around the nucleus of an atom.

Thus, there are likely pathways or channels created by the interplay of the protons and neutrons confined in the nucleus of an atom, interacting with the three-dimensional quadsitrons aether within and surrounding the boundaries of the atom. See Figure 55.

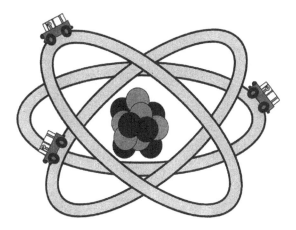

Figure 55

Pathways created by the interplay of protons and neutrons occupying the nucleus with the free quadsitrons within and surrounding the boundaries of the atom. The added imagery of the cars is meant to suggest the potential roadways (orbitals) always exist for the electrons to utilize.

Like interstate highways which exist for car travel, pathways for electron orbits may continuously exist for electrons as a result of the composite of the nucleus of an atom. The more protons and neutrons which exist in the nucleus of an atom, the more intense the structure applied to the free quadsitrons surrounding the nucleus and the greater number of orbitals exist for electrons to utilize when circling the nucleus of an atom.

CHAPTER 21

LAMNISCATE MODEL OF THE ATOM: RE-DEFINING THE STRUCTURE OF THE ATOM WITH THE NOVEL VERSION OF THE ELECTRON

THE TERM LAMNISCATE REFERS to decorating with ribbons. With regards to math, lamniscate is sometimes used to describe a series of figure-of-eight curves or curves which appear similar to the infinity symbol. The Lamniscate Model of the atom, therefore is rooted in the concept that electrons, within the structure of an atom, may follow figure-of-eight pathways around two or more protons. Neutrons balance the position of the protons in space within the atom. Electrons, occupying internal orbitals of an atom, weave around neutrons as they orbit the protons.

The structure of the atom remains an elusive mistress. As discussed earlier, at one time, the atom was thought to be comprised of a pudding like substance, which held negatively charged particles suspended in a matrix as proposed by Sir Joseph Thompson in 1904.[13] For a time, the atom was thought to be constructed similar to the construct of the solar system, with electrons circling the nucleus of an atom likened to planets orbiting the sun.

Niels Bohr, devised a model of the atom suggesting the electrons orbited about the center or nucleus of the atom in distinct orbitals. Electrons were capable of jumping between orbitals. Niels Bohr also described that the periodicity of the elements in the Periodic Table, were explained by the electron orbital structure of the atom.

There exist 118 known elements comprising the Periodic Table.[72, 73, 74] To construct these various elements there are nine shells and eight subshells. Originally the shells were referred to by letters, including: K, L, M, N, O, P, and Q. The current nine shells are referred to as: 1st shell, 2nd shell, 3rd shell, 4th shell, 5th shell, 6th shell, 7th shell, 8th shell and the 9th shell. The eight subshells are labeled: 's', 'p', 'd', 'f', 'g', 'h', 'I', and 'j' orbitals. The later shells (>7) and later subshells (>g) configurations become increasingly theoretical. The subshells fill by the Aufbau principle; in the ground state of an atom, the electrons fill atomic orbitals of the lowest energy levels before filling higher energy levels.[75] The first shell has only an s orbital. The second shell may have both s and p orbitals. The third shell may be comprised of s, p and d orbitals. The fourth shells may be comprised of s, p, d and f orbitals.

In the Electron Shell Model of an atom, the number of protons dictates which element the atom is and, by convention, the number of electrons an atom may have assigned to its orbitals to be considered stable regarding positive and negative charges. The number of protons, generally also dictates the number of neutrons occupying the nucleus aside the protons, though this statement is loosely adhered to by atomic structures; due to the number of neutrons being a stabilizing factor for an atom's three-dimensional physical structure. Isotopes refer to variants of an element where given the set number of protons and electrons, there exists a varying number of neutrons, either more or less than the number of protons present in the nucleus.[76]

The current Electron Shell Model of an atom was first established in 1923 by Niels Bohr and has undergone numerous revisions since first presented to the scientific community. The initial atomic model consisted of simple sphere-like orbitals, often drawn on a piece of paper as circles around a central nucleus. The Electron Shell Model of the atom has dominated the contemporary conscious thought and academic teaching regarding the structure of the atom. Arnold Sommerfeld (1868-1951) modified Bohr's model of the atom, by adding elliptical orbits to the original planetary model design.[16] Electron shells were later observed experimentally per x-ray absorption studies performed by Charles Barkla (1877-1944) and Henry Moseley (1887-1915).[77,78] Charles Barkla originally labeled the electron shells as K, L, M, N, O, P and Q; the letters K and L originally appeared in 1911 in the paper *The Spectrum of Fluorescent Rontgen Radiations* by Barkla. The electron shell nomenclature has matured to be currently numbered as: 1, 2, 3, 4, 5, 6, 7, 8, and 9; with subshells designated as: s, p, d, f, g, h, I, and j; again, with the later shells and subshells being more theoretical.

The current model of the atom sports rather elaborate appearing electron orbitals of differing shapes depending upon the subshell. The possible orbitals include 1s, 2s, 2p, 3s, 3p, 3d, 4s, 4p, 4d, 4f, 5s, 5p, 5d, 5f, 5g, 6s, 6p, 6d, 6f, 7s, 7p, 7d, 8s, 8p, and 9s. Though there is a uniform algorithm to the speculative design of the atom, the orbitals do not necessarily fill in the order of the above labeling. Filling behavior of electron orbitals in some cases may be element specific.

The valance electrons refer to the bonding electrons. Valance orbital tend to be the most distant orbitals from the nucleus of an atom. Valance electrons are capable of participating in chemical bonds. An atom's valance electron(s) are capable of being shared with other atoms, which can result in covalent bonds between atoms to create molecules. The one, two, three or four electrons which occupy the outer orbitals of atom are the workhorse of the atom. The electrons held in lower orbitals, closer to the nucleus of the atoms generally are held tight to the atom and do not participate in chemical bonding.

Erwin Madelung (1881-1972) developed the Madelung rule for the filling of atomic orbitals, which is as follows: the atomic orbitals are filled in order of increasing 'n + l' quantum numbers, where 'n' is the principle quantum number and 'l' is the azimuthal quantum number.[79] The azimuthal number determines the orbital angular momentum and shape of the orbital. The actual filling of the orbitals is described as follows: 1s, 2s, 2p, 3s, 3p, 4s, 3d, 4p, 5s, 4d, 5p, 6s, 4f, 5d, 6p, 7s, 5f, 6d, 7p, (8s, 5g, 6f, 7d, 8p, and 9s); noting that after the 7p orbital, the orbitals become less clear in the ground state of the heavier atoms. It is also noted that in addition to s, p, d, and f subshells, there may exist g, h, i and even j subshells. What is clear is the current version of the Electron Shell Model of the atom, still does not accurately explain how the spectrum of atoms are actually constructed or how the electrons within the boundaries of an atom, truly behave in the real world.

The electron shell configuration of an atom suggests an electron is an object of sorts, with a mass and travels at the speed of light. Electrons are capable of orbiting the nucleus of an atom, or hopping from atom to atom if pushed by a potential (voltage) applied across an electric circuit or electrons are thought to flow freely in some circumstances, such as flow from one electrode to another electrode in a cathode ray tube when the cathode is heated to an excitable state.

Again, this text takes the position that electrons are comprised of (1) a quanta of energy traveling at light speed and at any given instant in time (2) a set of 463 consecutive rings of quadsitrons physically surround the quanta of energy. Orbitals are positioned around the nucleus of an atom consisting of channels created by the arrangement of quadsitrons due to the forces exerted by the arrangement of protons and neutrons at the center of the atom. These channels are potential roadways for quanta of energy to flow through, an electron being a quanta of energy and 463 rings of quadsitrons a subset of quadsitrons of a channel (orbital). The orbitals are set up as the result of the three-dimensional combinations of arrangements of protons and neutrons occupying the nucleus of an atom, which exerts a force on the quadsitrons in three-dimensional space surrounding and occupying the interior of an atom. Such pathways are directly dependent upon the number of protons occupying the nucleus and such pathways exist even if an electron is not actively utilizing the pathway.

The Periodic table provides a summary of the 118 known elements which have been identified to date. The following is a table comparing some of the characteristics of several elements discussed in this text. As the characteristics of the elements are dissected, the assumptions used in the construct of the Electron Shell Model of the atom, which governs the current thought on the structure of atoms, comes into question.

ELEMENT NAME Atomic Weight ORBITALS	ABRV	ATOMIC NUMBER Protons/ Neutrons	PRIMARY STABLE ISOTOPE	SECOND STABLE ISOTOPE	ATOMIC RADII* Covalent RADII (pm)	Van der Waals RADII (pm)
Hydrogen 1.008 $1S^1$	H	1/0	^1H 99.98%	^2H 0.02%	53 31±5	120
Helium 4.002 $1S^2$	He	2/2	^4He 99.99%	^3He 0.0002%	31 28	140
Lithium 6.94 [He] $2S^1$	Li	3/4	^7Li 95%	^6Li 5%	167 128±7	182
Beryllium 9.012 [He] $2S^2$	Be	4/5	^9Be 100%	^7Be, ^{10}Be Trace	112 96±3	153

Boron 10.81 [He] $2S^2, 2P^1$	B	5/6	^{11}B 80%	^{10}B 20%	87 84±3	192
Carbon 12.011 [He] $2S^2, 2P^2$	C	6/6	^{12}C 98.9%	^{13}C 1.1%	67 sp^3 76±1 sp^2 73±2 sp 69±1	170
Nitrogen 14.007 [He] $2S^2, 2P^3$	N	7/7	^{14}N 99.6%	^{15}N 0.4%	56 71±1	155
Oxygen 15.999 [He] $2S^2, 2P^4$	O	8/8	^{16}O 99.76%	^{18}O 0.20%	48 66±2	152
Fluorine 18.998 [He] $2S^2, 2P^5$	F	9/10	^{19}F 100%	^{18}F Trace	42 57±3	147
Neon 20.179 [He] $2S^2, 2P^6$	Ne	10/10	^{20}Ne 90.48%	^{22}Ne 9.25%	38 58	154
Sodium 22.989 [Ne] $3S^1$	Na	11/12	^{23}Na 100%	^{22}Na, ^{24}Na Trace	190 166±9	227
Magnesium 24.305 [Ne] $3S^2$	Mg	12/12	^{24}Mg 79.0%	^{26}Mg 11.0%	145 141±7	173
Aluminum 26.981 [Ne] $3S^2 3P^1$	Al	13/14	^{27}Al 100%	^{26}Al Trace	118 121±4	184
Silicon 28.085 [Ne] $3S^2 3P^2$	Si	14/14	^{28}Si 92.2%	^{29}Si 4.7%	111 111±2	210
Phosphorous 30.973 [Ne] $3S^2 3P^3$	P	15/16	^{31}P 100%	^{32}P Trace	98 107±3	180
Sulfur 32.06 [Ne] $3S^2 3P^4$	S	16/16	^{32}S 94.99%	^{34}S 4.25%	88 105±3	180

Chlorine 35.45 [Ne] $3s^2 3p^5$	Cl	17/18	^{35}Cl 76%	^{37}Cl 24%	79 102±4	175
Argon 39.948 [Ne] $3s^2 3p^6$	Ar	18/22	^{40}Ar 99.604%	^{36}Ar 0.334%	71 106±10	188
Potassium 39.098 [Ne] $2s^2 2p^6 3d^1$	K	19/20	^{39}K 93.258%	^{41}K 6.73%	227 203±12	275
Copper 63.546 [Ar] $3d^{10} 4s^1$	Cu	29/34	^{63}Cu 69.15%	^{65}Cu 30.85%	121 132±4	140
Krypton 83.798 [Ar]$3d^{10} 4s^2 4p^6$	Kr	36/48	^{84}Kr 56.99%	^{86}Kr 17.28%	88 116±4	202
Xenon 131.293 [Kr]$4d^{10} 5s^2 5p^6$	Xe	54/78	^{132}Xe 26.909%	^{129}Xe 26.401%	108 140±9	216
Gold 196.966 [Xe] $4f^{14} 5d^{10} 6s^1$	Au	79/118	^{197}Au 100%	^{195}Au synthetic	174 136±6	166
Radon (222) nonstable [Xe] $4f^{14} 5d^{10}$ $6s^2 6p^6$	Rn	86/136	^{222}Rn Trace	^{211}Rn synthetic	120 150	220
Oganesson (294) nonstable [Rn] $5f^{14} 6d^{10} 7s^2$	Og	118/176	^{294}Og synthetic	^{295}Og synthetic	No data 157	No data

Atomic Weights [80],
Covalent radii: Cambridge Structural Data Base[81]
Atomic Radii of the elements (data page)[82]
Atomic Radii[83]
Synthetic = man made

Table 2

Characteristics of elements discussed in this text

As mentioned above and demonstrated in the table, the Electron Shell Model of the atom suggests there are orbitals positioned about the nucleus of an atom. Electrons occupy the orbitals, while protons and neutrons are positioned in the nucleus. As the number of protons increases in the nucleus of an atom, the number of electrons also increases generally at a ratio of 1:1. As the number of electrons orbitals increases to fill orbitals, additional electrons fill alternative orbitals.

There are several pressing concerns regarding the Electron Shell Model of the atom. Copper with an atomic number of 29 and gold with an atomic number of 79, both have a smaller radius than sodium with an atomic number of 11. One might conclude the smaller number of protons in the nucleus of a sodium atom are not capable of holding the $3S^1$ electron in as tight or close in orbit, as copper is able to hold its $4s^1$ electron or gold is able to hold its $6s^1$ electron. But then the reverse is true, as the nucleus of an atom, such as copper with 29 protons and 34 neutrons, and gold with 79 protons and 118 neutrons, vastly increases the number of protons and neutrons, it would seem logical the electrons occupying p, d, f orbitals would at some time crash into the nucleus of the atom or other electrons given the design of the orbitals.

The Electron Shell Model of atomic structure comes into question regarding the model's failure to provide an explanation as to the means as to why packing protons and neutrons in the nucleus creates electron orbitals around the nucleus. Points which require clarification include: (1) by what manner do 's' orbitals or 'p' orbitals or 'd' orbitals or 'f' orbitals form, given the number of protons occupying a nucleus. (2) why does adding a proton and two neutrons to helium atom to generate a lithium atom, create a $2S^1$ electron orbital. (3) why does adding a proton and two neutrons to the nucleus of a neon atom to generate sodium physically create the spherical $3S^1$ electron orbital. (4) why in many instances do the number of protons not match the number of neutrons in the nucleus. (5) what is the stabilizing factor neutrons provide atoms. (6) why, in the above table, are the number of neutrons generally an even number; is there a physical balance to the construct of the nucleus, which neutrons are directly responsible for given their presence.

There are many questions regarding the Electron Shell Model which need answering. To try to explore answers to the questions we will look at how atoms are constructed, starting with hydrogen. It is to be noted, that Nature has constructed all of the differing atoms, which exist in the Periodic table, without the benefit of design or formal manufacturing. The different elements exist in their structural forms due to 'stability' derived from some form of inherent acrobatic interaction between the protons and neutrons comprising the nucleus and the electrons occupying the orbitals of the atom.

HYDROGEN

Utilizing the teachings of the Electron Shell Model of atomic structuring, we consider the hydrogen (H) atom, where there exists one proton and one electron, there is one orbital, designated as $1s^1$. See Figure 56(a). There is no neutron in a hydrogen atom. The hydrogen atom has an atomic number of 1, an atomic weight of 1.008. Thus, the occurrence of an electron orbital is not dependent upon the presence of neutron. In the hydrogen atom, the single electron is believed to travel within the outer boundaries of a three-dimensional sphere around the proton at the center of the atom. The 's' orbitals of an atom are believed to be in the shape of a sphere. The electron comprising a hydrogen atom is very amendable to combining with other atoms which are seeking electrons to balance their orbitals in what is termed covalent bonding, also termed a shared bonding of electrons between atoms.

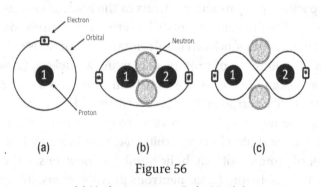

(a)   (b)   (c)

Figure 56

(a) Hydrogen atom, Bohr Model,
(b) Helium atom spherical orbital, Electron Shell Model,
(c) Helium atom with an alternative figure 8 orbital, Lamniscate Model.

HELIUM

Helium (He) is an atom comprised of two protons, two neutrons and two electrons. Helium has an atomic number of 2, atomic weight of 4.0026. The Electron Shell Model approach to the helium atoms is seen in Figure 56(b). The 's' shell is capable of accommodating two electrons, thus the spherical s-orbitals can either have one or two electrons present in the orbital. The construct of helium fits the Electron Shell Model of an atom very nicely.

Since the 1s orbital is full in a helium atom, the electrons of a helium atom are not amendable to be shared with other atoms. Helium is considered to be an inert element. Helium atoms are generally found in nature as isolated atoms, not combined with other atoms. Given earth's environment, helium is generally found as a gas.

The concept that the 1s shell is a sphere is simple to conceptualize, since it is easy to consider a planet orbiting the sun exhibits a circular-like, almost spherical process. But two electrons traveling in a spherical orbit around the nucleus of a helium atom does not really explain why the two electrons are impotent in their action to combine with other atoms. One logical explanation is that these two electrons are being held tightly to the nucleus with a shorter radius, more so than other electrons that would orbit the nucleus.

An alternative theory would be that the two electrons in the 1s orbital travel through the center of the atom. See figure 56(c). If the energy of the electrons pass through the center of the atom, then it is logical that the electrons are held tightly bound to the atom. Electrons which would travel through the center of the atom as part of the course of the electron orbital would have less opportunity to be coaxed away from the nucleus of the atom and participate in a covalent bond with another adjacent atom.

To this point the first two elements of the 118 elements comprising the Periodic Table of Chemistry has been discussed. As described thus far, the scientific explanation of the existence of the modern structure of an atom's s, p, d and f shells holds somewhat true. Two electrons comprising the first 's' shell configuration has not significantly challenged the logic of analysis.

LITHIUM

The third element is lithium (Li). The number three type of atom is generally considered to be comprised of three protons, three neutrons and three electrons. Lithium has an atomic number of 3, atomic weight of 6.94. Electron Shell Model of the Lithium atom seen in Figure 57(a).

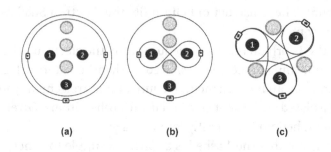

(a) (b) (c)

Figure 57

(a) Electron Shell Model Lithium atom,
(b) Lithium atom with figure 8 for 1s orbitals and spherical orbital for 2s,
(c) Lithium atom with equal figure 8 orbital for all three orbitals.

Lithium has several potential isotopes. The most abundant form of lithium is Li-7, which is comprised of three protons, four neutrons and three electrons. Li-7 comprises 92.5% of all naturally occurring lithium. Approximately 8% of lithium is found as Li-6.

The Electron Shell Model of the lithium atom would assume the first two protons remain tightly held together as the third proton is added. Logic suggests that as the third proton is brought into the construct of the nucleus, there would occur a conformational change in position of the first two protons. The change in the position of the first two protons, by the appearance of the third proton, must in some way alter the shape of the orbital for the two electrons occupying the 1s orbital. There must somehow be a balance in the shape of the Lithium atom's nucleus. Conceivably, the third proton could simply revolve around the axis created by the position of the first two protons. See Figure 57(b). Alternatively, the three protons could form a triangular shaped structure, which logically would be the most stable structure. See Figure 57(c).

Lithium tends to bind with one other atom. Lithium carbonate is a medication commonly used to manage bipolar disease. Figures 57(a) and

57(b) lend themselves to explaining why Lithium is capable of binding with one other atom, suggesting that the existence of the s orbital allows for that electron to covalently bond with another atom. The lithium atom configuration illustrated in Figure 57(c), though suggesting a more logical balance of the three protons in space, does not provide a sense that there exists a free electron to participate in the process of covalent bonding.

Taking into consideration the fourth neutron seen in the most abundant form of lithium, the Li-7 isotope, does not appear to shed any light logic behind the structure of the lithium atom and the covalent bonding capacity of the lithium atom. See Figure 58. Adding a fourth neutron to the nucleus, structurally appears to stabilize the atom in all three illustrations to include (a), (b) and (c). The behavior of the atom and the structure of the atom suggests the third proton rotates within the boundaries of the nucleus around the axis created by the first two protons. Such a rotation of the third proton would suggest a wobble in the shape of the second s shell orbital should occur as the third proton moves about the axis of the nucleus.

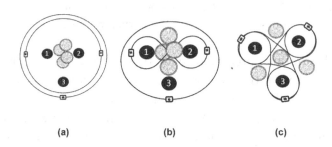

(a) (b) (c)

Figure 58

(a) Electron Shell Model Lithium-7 isotope,
(b) Lithium-7 isotope with figure 8 for 1s orbitals
and wobbling spherical orbital for 2s,
(c) Lithium-7 isotope with equal figure 8 orbital for all three orbitals.

Thus far the Electron Shell Model for atomic structures continues to properly represent the construct of how protons, neutrons and electrons are theoretically arranged in the construct of atoms.

BERYLLIUM

Beryllium (Be) is the fourth element. The atomic number is 4. Atomic weight is 9.0121. Beryllium is considered to have four protons, four neutrons, and four electrons. Beryllium has four isotopes, but only one is considered stable. Nearly all Beryllium is found as the Beryllium-9 (Be-9) isotope, which contains four protons, five neutrons and four electrons. See Figure 59.

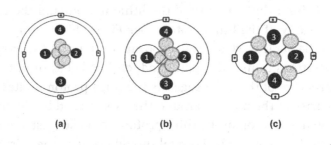

(a) (b) (c)

Figure 59

(a) Electron Shell Model for Beryllium-9 isotope,
(b) Figure 8 orbital for 1s orbital model for Beryllium-9 isotope,
(c) Lamniscate model for the Beryllium-9 isotope.

The illustrations presented for both 59(a) and 59(b) representing possible versions of the Be-9 isotope continue to be consistent with the Electron Shell Model for the atom. Illustration 59(c) represents a Lamniscate Model of the Be-9 isotope showing potential positions of the neutrons and protons and compact construct of the orbits, which keeps all four electrons tight to the nucleus, preventing bonding of Be-9 isotope's electrons with other atoms.

BORON

The Boron atom challenges the Electron Shell Model of the atom. Boron (B) has an atomic mass of 5. Boron has an atomic weight of 10.81. Boron has two predominant isotopes, that of B-10 found 20% of the time and B-11 found 80% of the time. B-11 has five protons, six neutrons and five electrons. See Figure 60.

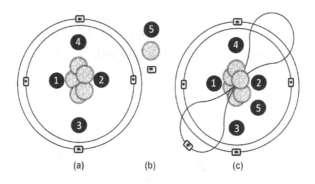

Figure 60

(a) Beryllium-9 Electron Shell Model with the addition
of (b) a proton, one neutron and one electron to create
(c) Electron Shell Model Boron-11 isotope.

The Boron Electron Shell Model as shown in Figure 60 (c) suggests that to generate a Boron-11 isotope, one proton, one neutron and an electron are added to a Beryllium-9 isotope. See Figures 60(a) and 60(b). Where the Beryllium atom appears rather symmetrical and balanced, the concern is the addition of a fifth proton creates asymmetry to the Beryllium atom. The additional concern is that now a dumbbell like orbital is now created by adding a proton and neutron to the structure. See Figure 60 (c). How this dumbbell like orbital is contrived is unclear per the Electron Shell Model of the atom. The Electron Shell Model for an atom is based on filling orbitals due to an effort by atoms to migrate to filling orbitals to satisfy balances of positive and negative charges. Hydrogen and helium are stable if there are two electrons to complete their orbitals. Lithium and Beryllium are thought to be complete if there are four electrons in their orbitals. Atoms with atomic numbers 5-9, which include boron, carbon, nitrogen, oxygen, and fluorine are thought to actively seek to fill their orbitals until there is a total of ten electrons comprising the electron cloud about the atom's nucleus.

Curiously, if one were to entertain the thought that the addition of the fifth proton to the Beryllium-9 isotope would cause a balanced rearrangement of the protons and neutrons within the nucleus, then pentagon like structure seen in Figure 61(a) might be possible. Alternatively, if the idea is maintained that the first two protons remain bound tightly to each other, then the remaining three protons are allowed to position themselves around the center axis of the first two protons as seen in Figure 61(b) and 61(c). Boron tends to bind to three other atoms by covalently bonding three electrons. In special circumstances, Boron is capable of binding a total of five additional electrons.

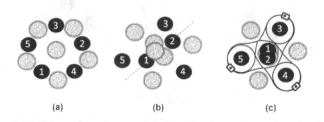

(a)   (b)   (c)

Figure 61

(a) Boron-11 isotope with a pentagon shaped nucleus,
(b) Side view of Boron-11 isotope with three protons about a center axis,
(c) Top view Boron-11 isotope with Figure 8 orbitals, Lamniscate Model.

The structure of the B-11 isotope as illustrated in Figures 61(b) and 61(c) suggest that three orbitals could exist, given the number of protons present in the nucleus, which could lend an electron to covalently bond to other atoms.

CARBON

A point of confusion regarding the position of the Electron Shell Model's orbitals is explaining how an atom such as 'carbon' is built, and then how such an atomic structure functions. Carbon (C) has an atomic number of 6. Atomic weight of carbon is 12.011. Carbon is thought to be the key element in the construct of most organic structures. Carbon-12 is the most abundant isotope accounting for 98.9% of the available carbon, is comprised of six protons, six neutrons and six electrons. Carbon-13 comprises 1.1% of carbon atoms, while Carbon-14, often used in the process of carbon dating by archeologists, is relatively rare.

In the Electron Shell Model, the six electrons of a carbon atom are thought to occupy the 1s, 2s and 2p shells around the nucleus of the atom. Two electrons are thought to occupy the 1s shell. Two electrons are thought to occupy the 2s orbitals. Two electrons are thought to occupy the 2p orbitals. See Figure 62. Again, carbon has six protons. The carbon atom is thought to have four empty orbitals in the 2p orbitals. The four empty orbitals are thought to allow the carbon atom the potential to covalently bond to up to four atoms. One example would be the methane molecule, where one carbon atom is bonded to four hydrogen atoms. Many organic molecules are comprised of lengthy carbon atom chains where a carbon atom is covalently bonded to a carbon atom on either side and two hydrogen atoms. The unexplained phenomenon, is how an atom with six protons has the construct to bond to ten electrons where in effect, it should only be interacting with six electrons. The obvious question is, where does the subatomic scaffolding structure come from for the carbon atom to bind to four additional electrons?

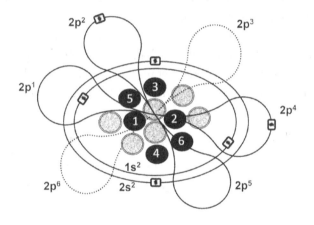

Figure 62

Illustration of an Electron Shell Model of the carbon
atom showing a 1s, 2s and 2p orbitals.

The oddity posed by the carbon atom is that two electrons are present in both the 1s and 2s shells, leaving two electrons to occupy the six p-orbitals. The shorthand notion for the carbon atom is $1s^2, 2s^2, 2p^{1,2,0,0,0,0}$. As shown in Figure 62, one p-orbital is illustrated with broken lines and no electron occupying the orbital, to draw attention to the fact that there is no practical reason as to why this orbital should exist. It is plausible that protons 5 and 6, interacting with the two p-orbital electrons creates the dumbbell shaped p-orbitals, but logic would suggest that at most there are only enough atomic pieces to create two p-orbitals; then where does the third dumbbell shaped p-orbital come from; where is the substance to create such an orbital? This is physics, not a theatrical production (where certain liberties can be taken); there should be both a well-grounded substantial explanation, as well as a mathematical modeling explanation for this, the most fundamental observance of atomic construction. In addition, atomic structures must form as a result of pure natural and random circumstances related solely to the principles of physics, without any outside assistance in their construct, such as that which might be afforded by intervention by a designer. That is, the construct of atoms has to have been a process which occurred purely on its own.

In studying the available pieces to the atomic puzzle, again there are six protons. If the first two protons (1 and 2) are considered to have a special bonding and are held tighter together in comparison to the remaining protons, this partitions the remaining four protons and allows for separate bonding to occur. See Figure 63(a). In three-dimensional space the carbon atom can be very balanced. Alienated from the first two protons, protons 3 and 4 may set up an exclusive orbital and likewise protons 5 and 6 may set up an exclusive orbital. See Figure 63 (b), (c), (d).

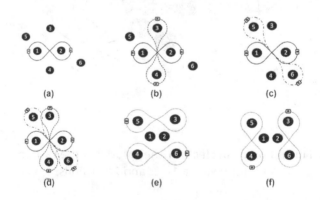

Figure 63

Carbon atom: (a) six protons with first two in a figure 8 orbital, (b) protons 3 and 4 create an orbital, (c) protons 5 and 6 create an orbital, (d) three orbitals shown which pass through the center of the atom, (e) protons 3 and 5 & 4 and 6 share exterior orbitals, (f) protons 5 and 4 & 3 and 6 share exterior orbitals.

An alternative means of studying the existing structures, would be to redesign the pathways utilized by the orbitals. In the original model with six protons and two dumbbell shaped orbitals with the electrons passing through the center of the atom, logic would dictate that the carbon atom orbitals should be full, there would be no additional bonding, unless additional orbitals are fabricated. Alternatively, if one imagines exterior orbitals, where the electron pathways do not pass through the center of the atom, as shown in Figure 63 (e) and 63(f), and consider that more than one electron may occupy the same space, then given the construction constraints eight orbitals are possible.

Adding the neutrons to the illustration of the carbon atom nucleus shows even further stabilization of the carbon atom structure, by the neutrons. The addition of the neutrons as depicted in the illustration, suggest that exterior pathways for the p-orbitals may indeed be encouraged by the construct of the atomic structure. Note when viewing the carbon atom in Figure 64, one needs to think of the special architecture of the atom in three-dimensional space, rather than a two-dimensional picture. The depiction of a carbon atom in Figure 64 suggests that there would be a 1s shell and 2p orbitals, but no 2s orbitals. All orbitals in the second shell would be p-orbitals.

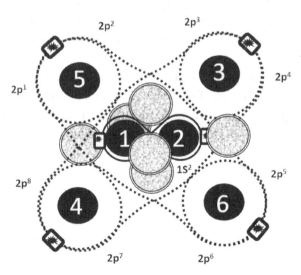

Figure 64

Carbon-12 atom illustrated with six protons, six neutrons and six electrons utilizing a Lamniscate Model.

Thus, the Electron Shell Model of an atom, as drawn for a carbon atom, would suggest the presence of two spherical shaped s-orbitals, with two electrons occupying both s-orbitals, and then two electrons occupying two of six possible dumbbell shaped p-orbitals. The short-hand description for this would be $1s^2, 2s^2, 2p^2$. See Figure 65 (a). A carbon atom with an equal set of electron orbitals external to the center of the atom would be considered to have eight electrons in the 2p-orbitals rather than six as suggested by the Electron Shell Model. The short-hand description for this would be $1s^2, 2p^4$ or possibly $1s^2, 2p^{4,0,0,0,0}$. See Figure 65 (b).

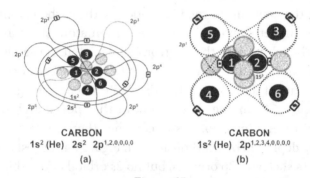

CARBON
$1s^2$ (He) $2s^2$ $2p^{1,2,0,0,0,0}$

(a)

CARBON
$1s^2$ (He) $2p^{1,2,3,4,0,0,0,0}$

(b)

Figure 65

Carbon-12 atom: (a) Electron Shell Model with p-orbitals crossing through the center of the atom, (b) Lamniscate Model illustrated with p-orbitals being exterior to the center of the atom.

The Lamniscate Model of the carbon-12 atom suggests that there are four outer protons and four outer electrons in a stable carbon atom. Given the figure eight pathway around the protons, the four outer protons of the carbon atom can accept and share bonding with four additional electrons due to the eight potential orbital pathways which surround the outer protons. This model has the appearance of a stable atomic structure.

NEON

Larger atoms become even more confusing as to how the nucleus is constructed. The interaction between the protons and neutrons is vitally important with regards to how the nucleus of an atom interacts with the electron orbitals of the same atom. Neon (Ne) is an inert element. Neon's atomic number is 10. Neon's atomic weight is 20.179. Neon-20 isotope accounts for 90.48% of the neon atoms. Neon-21 isotope accounts for 0.27% of the neon. Neon-22 isotope accounts for 9.25% of the stable neon atoms. The Electron Shell Model of the Neon atom would suggest that the $1s^2$, $2s^2$ and $2p^6$ orbitals are all occupied by electrons. See Figure 66. That the neon atom is not seeking to accept any additional electrons, nor is the atom looking to release any of its ten electrons to another atom.

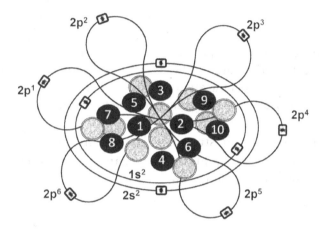

Figure 66

Neon-20 atom as illustrated by Electron Shell Model protocol.

The curious thing is by what mechanism are the 10 electrons in orbit around the nucleus of the neon atom held so tightly that they are not amendable to being shared with other atoms. The accepted reason is that the 1s, 2s and 3p orbitals are all filled and therefore the atom is in balance and has no propensity to interact with other atoms. What is unclear is the exact structural reason as to why the atom is in balance. Study of the positioning of the protons and neutrons in the nucleus may arrive at the answer. Possibly the protons are surrounded by the neutrons, pushing the protons closer together in the nucleus. See Figure 67. The action of positioning the protons tighter in the nucleus may result in making the electron orbitals closer in relation to the nucleus, even causing the orbitals to pass through the center of the atom, making the electrons less interactive.

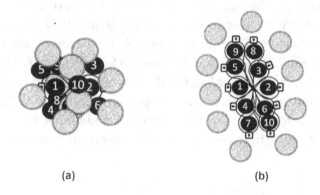

(a)   (b)

Figure 67

Neon-20 isotope: (a) nucleus with protons crowded around
the nucleus of the atom, (b) protons packed tightly together
in the nucleus of the atom with figure 8 orbital, making
the orbitals less accessible for electron bonding.

More than likely the illustration in Figure 67 (b) is not representative
of a neon atom. But expounding upon the Lamniscate Model, Figure 68
could represent a stable neon atom.

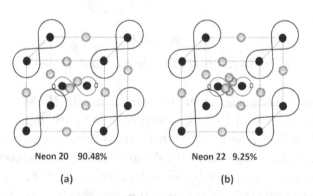

Neon 20   90.48%   Neon 22   9.25%

(a)   (b)

Figure 68

Lamniscate Model of the Neon atom, illustrating how the figure 8 orbitals
could be represented as part of an atomic structure. (a) Neon 20. (b)
Neon 22, notice two additional neutrons occupy the center of the atom.

The Lamniscate Model construct of the Neon atom is that of a
cube-like entity with the neutrons positioned to stabilize the structure
of the protons as the protons interact with the electrons. The presence

of neutrons within the cube-like structure direct electron pathways as well as restrict locations were electrons may orbit within the atom.

SODIUM

The generation of a sodium atom is derived from adding a single proton and electron to the neon atom. Neon's atomic number is 10, sodium's atomic number is 11. One hundred percent of naturally found sodium has 11 protons and 12 neutrons. The atomic weight of sodium is 22.989. Thus, by the Electron Shell Model of the atom, by adding a proton and two neutrons, the structure of the neon atom changes to include the $3s^1$ orbital. It is unclear how the massing of 11 protons and 12 neutrons in the nucleus of a sodium atom results in the construct of the multiple 's' orbitals and 'p' orbitals. See Figure 69.

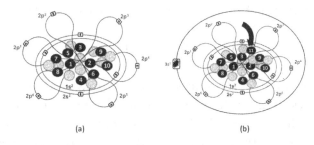

(a)   (b)

Figure 69

Electron Shell Model of the atom, how does adding one proton to Neon, create the 3S1 orbital required to produce the Sodium atom.

What structural changes at the subatomic level adequately explain the phenomena that adding one proton and two neutrons creates the a spherical 3s orbital as the atomic structure morphs from a Neon atom to a Sodium atom? An alternative to the Electron Shell Model of the atom for sodium would be the Lamniscate Model. See Figure 70. The Lamniscate Model, with figure eight appearing electron orbitals, suggests a stable atomic structure for an atomic structure with eleven protons.

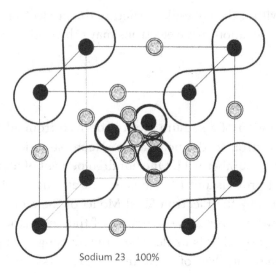

Sodium 23   100%

Figure 70

Lamniscate Model of the Sodium atom, illustrating how the figure
8 orbitals could be represented as part of an atomic structure.

Atoms may be dynamic, rather than static structures, thus the core
of the atom may actively revolve around an axis. The tumbling of the
three protons at the center of the sodium atom may produce unbalanced
physical forces on the protons and electrons occupying the outer aspects
of the atom. The unbalance in the outer protons and electrons caused by
the repositioning of the core of the atom, may result in, at times, one of
the peripheral electrons being pushed further out of the bounds of the
atom, and thus this more distant electron being capable of participating
in covalent bonding. The sodium atom generally participates in one
covalent bond with one other atom.

ARGON

Argon (Ar) has an atomic number of 18, with atomic weight of
39.948. ^{40}Ar makes up 99.604% of stable argon, with 18 protons and
22 neutrons. Per the Electron Shell Model of the atom, argon would
have 1s, 2s, 2p, 3s and the 3p electron orbitals filled. See Figure 71 (a).
The addition of one proton and the loss of two neutrons generates a
potassium atom.

POTASSIUM

Potassium (K) has an atomic number of 19, with an atomic weight of 39.0983. ^{39}K makes up 93.258% of stable potassium, with 19 protons and 20 neutrons. Per the Electron Shell Model of the atom, potassium would have 1s, 2s, 2p, 3s and the 3p electron orbitals filled and one electron in a 3d orbital. See figure 71 (b). ^{39}K gains one proton, but loses two neutrons in the comparison to the ^{40}Ar.

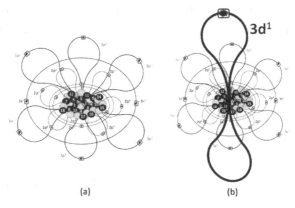

(a)   (b)

Figure 71

(a) Electron Shell Model of Argon atom. (b) Electron Shell Model of Potassium atom. Observation: how does adding one proton to Argon, create the 3d^1 orbital required to produce the Potassium atom.

The 'physical mechanism' remains unclear as to how 'adding' one proton and 'removing' two neutrons from an argon atom results in the generation of the construct of the potassium atom with a new '3d' orbital. In the case of the potassium atom and other large atoms, the Electron Shell Model of the atom represents science taking on a more mystical philosophy. The physics underpinning the structural changes in the atoms of the Periodic Table needs further clarification.

The question arises, is there an alternative structural design for atoms, which could be constructed based off of a simple, nonintelligent design, which could account for the complex structures of the 118 elements comprising the Periodic Table.

Potassium, with 19 protons and 20 neutrons. The Lamniscate Model of the Atom suggests the construct of a potassium atom could be comprised of a core component of five protons and electrons with an external component constructed of fourteen protons and electrons. See Figure 72. The neutrons could be distributed within the atom's structure to stabilize the position of the protons. The protons could be linked to the electrons in arrangements of two protons with figure 8 orbitals. As mentioned for sodium, the five protons comprising the core of the atom may tumble actively within the atom resulting in imbalances in the position of the peripheral electrons. Rotation of the five core protons may result intermittently with an electron being pushed further from the center, with respect to the other 13 electrons and thus this most distant electron may be more amendable to covalent bonding with another atom.

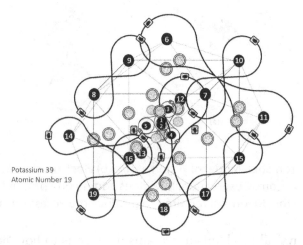

Potassium 39
Atomic Number 19

Figure 72

Lamniscate Model of Potassium atom

Presented here is an atomic structure consisting of 19 protons and electrons. In the Electron Shell Model, as the atomic structures become increasingly more complex, how do electrons in atomic orbitals not crash into each other or wreak havoc on and atoms internal structures. The explanation has been forthcoming for over a hundred years.

In the Lamniscate Model electrons have defined figure 8 pathways filled with one or two electrons, which generally do not interfere with the pathway of other electrons. Within each figure 8 orbital one electron moves with a right-handed spin, while the opposing electron moves with a left-handed spin; thus, if two electrons occupy a figure 8 orbital the electrons do not collide with each other. As the atomic structure becomes increasingly more complex by adding protons and electrons, with the Lamniscate Model design, the orbitals dynamically change to primarily insure stability of the atomic structure. Shells form in the Lamniscate Model, but changes may be more prominent within the internal structure of an atom rather than the outer perimeter of an atom's structure as seen with the Electron Shell Model. Stability of atomic structures is key, as is functionality of the electron orbitals. Increasing complexity of atomic structures threatens stability and functionality in the Electron Shell Model of an atom.

One must consider how the bulk of the universe's atoms have been created. How does adding a proton change the physical structure of an atom. How does adding one or more neutrons change the physical structure as well as contribute to the stability of the atom.

The Plumb Pudding Model of the structure of the atom was introduced by J.J. Thompson in 1904. Thompson wrote 'atoms of elements consist of a number of negatively electrified corpuscles enclosed in a sphere of uniform positive electrification'.[13] The term 'corpuscle' was an alternate expression for 'electron' at the time. The Plumb Pudding Model may be closer to the true representation of the structure of the atom, than the orbital patterns of the current Electron Shell Model of the atom.

What is clear, is that further investigation and discussion of the architecture of the atomic nucleus is a necessity. The Electron Shell Model of the atom as a structural representation of the atom is in serious trouble regarding validation of the nucleus and orbital forms from a physics standpoint. It is beyond the scope of this text to fully explore how the structuring of protons and neutrons in the nucleus of atoms affects the electron orbital architecture for all elements. What this chapter hoped to accomplish was to shine light on the fact that the understanding of electron orbitals requires further development and that the current understanding of how an electron and its orbital are constructed would benefit from further study, modification and update.

CHAPTER 22

WIRE, CURRENT, AND THE MAGNETIC FIELD

COPPER WIRE IS COMPRISED of copper atoms. As with any substance, there exist impurities, other types of atoms, in copper wire, but the majority of atoms would be copper atoms. As illustrated in Figure 73, copper atoms are constructed such that the electron in the 4s orbital can be caused to jump from one copper atom to an adjacent copper atom. The purer the copper content of the wire, the higher the wire's property of conductivity.

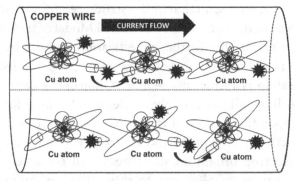

Figure 73

Copper atoms are constructed such that the electron in the 4s orbital can be caused to jump from one copper atom to an adjacent copper atom.

When a copper wire exists inert, in a state were no outside force is acting on the wire, there is no movement of the electrons through the copper wire; that is there is no movement of energy through the wire. When an energy potential is applied to the ends of a copper wire, where a source of energy is connected to a copper wire with a path for the energy to drain out of the wire, current flows through the wire. If electrons are indeed pathways set up by organized configurations of quadsitrons created by the forces at the center of the atom derived from the protons and neutrons, then the pathways for quanta of energy to pass from one copper atom to another copper atom are already in place to facilitate the flow of current. There has been entertained that the flow of energy through a circuit, referred to as current, is actually a flow of holes from copper atom to copper atom, rather than the traditional view that current is a flow of electrons. A model, where electrons are actually potential pathways defined by a segment of 463 consecutive rings of quadsitrons helps to substantiate the theory that the flow of electrons might be a flow of holes, or in other words, the use of the holes (pathways) which involve the 4s orbital of copper atoms.

As energy moves through a copper wire it is observed that a magnetic field moves along the same wire, but is oriented ninety degrees to the direction of the current flow. See Figure 74.

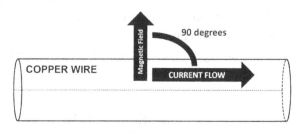

Figure 74

In a copper wire a magnetic field moves along the wire with a current, but oriented ninety-degrees to the direction of the current flow.

The concept of a magnetic field has been elusive, as mysterious as attempting to conceptualize the electron. Our senses tell us that electrons exist and that magnetic fields exist. For a century we have used magnetic fields to harness energy. Modern dams have often been designed to harness the flow of water to run turbines to create electricity by turning magnets inside of coiled wire. A magnet moving through the center of a coiled wire creates electron flow in the coiled wire.

If, in a copper wire, a potential is applied to the wire to cause a current to flow from one end of the copper wire to the opposite end of the copper wire, then quanta of energy are moving from one copper atom to the next copper atom in succession as long as the potential is attached to the wire and there is a reasonable resistance in the circuit. Resistance in a wire circuit would be opposition to flow of quanta energy; the ultimate resistance being an 'open' switch or dislodged wire connection, through which no electron could pass. Too high of a resistance in a wire circuit and the flow of the current will cease. Resistance in a wire circuit varies with amount of nonconductive impurities comprising the electron pathway.

It is conceivable that as a single quantum of energy, which a quanta of energy is theorized by quantum mechanics to be a unit of energy that is nonreducible, travels from the 4s orbital of one copper atom, to the 4s orbital of an adjacent copper atom the quanta of energy changes shape. As a quanta of energy leaves the confines of the 4s orbital of one copper atom and transfers to an open 4s orbital of an adjacent copper atom the quanta physically spreads out like a pancake. The quanta of energy may disperse out in the direction which is 90 degrees from the direction of travel the quanta of energy is taking from one copper atom to the next copper atom. If indeed the quanta of energy itself has changed shape, spreading its volume perpendicular to the axis of travel 360 degrees around the wire, this may represent what has been identified as the magnetic field surrounding a copper wire which has a current running down the length of the wire. See Figure 75.

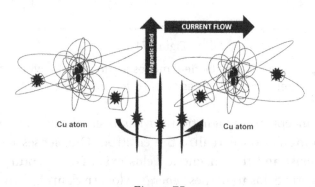

Figure 75

A magnetic field forms about a copper wire oriented
ninety degrees to the direction of the current flow.

Given that the magnetic field may be formed from quanta of energy traveling as current down the length of a copper wire; if as the quanta of energy spreads its volume ninety-degrees from the axis of current flow as the quanta transfers from one copper atom to an adjacent copper atom and encounters a substance which exhibits less resistance than the resistance present in the copper wire, then the quanta of energy may leap from the copper wire to the surrounding substance. This phenomenon would explain behavior of the movement of copper wire versus a magnet and the transfer of energy from copper wire to a magnet and vice versa. See Figure 76.

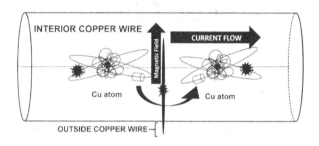

Figure 76

As one quanta of energy transfers from the 4s-orbital of one copper atom to the 4s-orbital of another copper atom, the quanta spreads its shape out like a pancake ninety degrees from direction of current flow and this is detectable as a magnetic field.

The function of the electric generator and the electric motor are based on utilizing a magnetic field to transfer electricity from one form to another form. In the case of the electric generator, often a natural phenomenon causes a rotation of a magnet, the rotation of the magnet causes electricity to be created inside a coiled wire surrounding the rotating magnet. In the case of an electric motor, electricity is run through a coiled wire, the coiled wire surrounds a rotatable magnet, as the current passes through the wire the magnet rotates about its axis.

The concept is that energy can be transferred from a moving magnet to a coiled wire. The general explanation for this phenomenon is that the magnet exerts a magnetic field and this magnetic field creates electricity in the wire in the case of a generator or alternatively a current flowing in a wire is capable of moving magnet positioned inside the coils of the wire. This concept has been analyzed, measured, and utilized to produce powerhouses to generate electricity or electric motors to run vehicles.

CHAPTER 23

REVISITING THE ELECTROMAGNETIC SPECTRUM

THE ELECTROMAGNETIC SPECTRUM IS the means by which both engineering and physics represents and explains how energy, which travels in waves, exists in the universe. Humans are most cognizant of the electromagnetic spectrum with our utilization of visible light to assist us in perceiving the environment which surrounds us. The concept of visible light represents the narrow band of wave energy that exists in a frequency range of 430 to 770 terahertz.[5] Other forms of wave energy exist, traveling the universe utilizing other frequencies, but the human eye is generally not adapted to view energy outside the visible light spectrum. To the human eye, frequencies of energy that exist outside the bandwidth of the visible light spectrum are invisible. Bees, many other insects, and some birds can see ultraviolet light, which human cannot. Pit vipers have a pair of infrared sensors, which assist them in detecting heat. Bats utilize echolocation or bio sonar, energy in the sound bandwidth in order to navigate at night.

The wave energy present in the electromagnetic spectrum is thought to travel from one location in the universe to another location in the universe in the shape of a sine wave. Frequency is considered the number of cycles of waves which occur in the period of a second. Wavelength is considered the distance between the maximum height of two sine waves. Given a sine wave, mathematically frequency is the inverse of wavelength. The speed of wave energy comprising the Electromagnetic Spectrum at any frequency is considered to be light speed in a vacuum.

The color white represents all of the visible light spectrum striking the human eye at the same time. Black represents the absence of visible light. The colors of light in increasing frequency include RED (400-484 Terahertz (THz)), ORANGE (484-508 THz), YELLOW ((508-526 THz), GREEN (526-606 THz), BLUE (606-668 THz), and VIOLET (668-789 THz). The spectrum of colors is continuous with no specific boundaries.[84] The retina of the eye absorbs visible light, sends the signals down the optic nerve from the back of the eyes to the posterior portion of the brain, where light images are interpreted in the occipital lobes of the brain. See Figure 77.

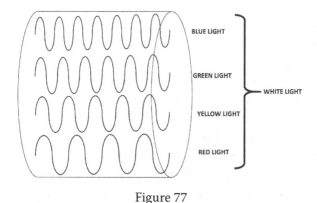

Figure 77

Illustration of varying wavelengths of visible light.

Visible light makes up a small, very specific portion of the overall electromagnetic spectrum.[85] See Figure 78. Larger wavelengths are used in power transfer and radio waves. The shortest wavelengths are represented by gamma rays. Gamma rays are considered to have the highest energy due to such waves exhibiting the maximum cycles per second.

Gamma rays have a frequency of approximately 300 EHz and a wavelength of 1 picometer. Hard x-rays have a frequency of 30 EHz and a wavelength of 10 picometers. Near ultraviolet is approximately 3 PHz, wavelength of 10 nanometers. Visible light as mentioned above, frequency 789-400 THz, with wavelength of micrometers. Infrared has a frequency of approximately 30 THz, wavelength of 10 micrometers. Microwaves have a frequency of 300 MHz, wavelength of 1 meter. Radio waves 3 MHz to 30 kHz, wavelengths of 100 meters to 10 km.

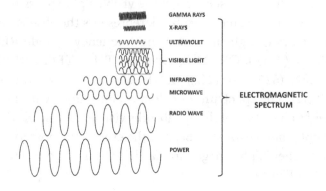

Figure 78

Visible light occupies a small bandwidth of energy compared to the entire bandwidth of energy comprising the electromagnetic spectrum.

Energy has been considered to travel from one location in space to an alternative location in space without the presence or assistance of a transfer medium. How an energy wave traveling as a sine wave is created and maintained without the assistance of an aether has remained a mystery; this is a concept that has been observed in a multitude of ways, without a valid explanation having been derived.

Wave energy traveling from one location to a second location in the universe has been identified as having two components. The two components have been described as electric energy and magnetic energy. The electric energy wave is considered to always be ninety degrees to the magnetic energy wave in relation to the axis of travel of the two waves.[86] See Figure 79.

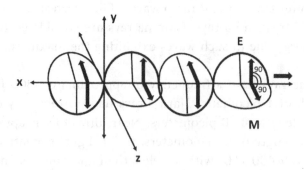

Figure 79

Electric wave energy (E) is considered always to be ninety-degrees to the magnetic wave energy (M) in relation to the axis of travel.

Alternatively, the construct of wave energy having to have a component of electric wave and positioned ninety degrees to the electric wave is a magnetic wave, without the presence of an aether is difficult to explain. If on the other hand, there exists a transfer medium comprised of quadsitrons, then a wave of energy can take on the shape of a three-dimensional object such as a sphere or series of spheres in three-dimensional space and the presence of both an electric wave and magnetic wave can be explained. See Figure 80.

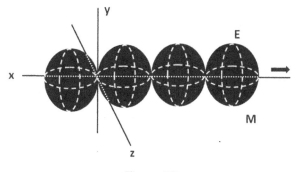

Figure 80

Wave energy can take on the shape of a three-dimensional object in three-dimensional space, that is energy can be a three-dimensional wave with both electrical component and the magnetic component oriented ninety degrees to each other and traveling in the same direction.

It is difficult to explain the behavior of wave energy across the frequencies of the electromagnetic spectrum without the presence of an aether or some form of underlying substance comprising the fabric of the universe. In three-dimensional space and accounting for a fabric of space which extends in all directions to the very edges of the universe, energy can travel from one location to another location as a three-dimensional wave. Such a three-dimensional wave has previously been described as part being electrical energy and part being magnetic energy.

QUADSITRON MECHANICS: APPLICATION TO THE PHYSICAL WORLD WHICH SURROUNDS US

CHAPTER 24

SOLVING THE MYSTERY OF THE BAR MAGNET

WE HAVE BEEN TAUGHT that a bar magnet is comprised of a positive end and a negative end, or that there is a north pole and a south pole. We are taught that like ends of a bar magnet repel each other, while opposing ends of bar magnets attract each other. That is, if the positive ends of two bar magnets or the negative ends of two bar magnets are placed adjacent to each other, then the poles will repel each other. Conversely, if the positive pole of one bar magnet is placed next to the negative end of a second bar magnet, then the two bar magnets will attract each other, and if in close enough proximity, the two bar magnets will physically breach the gap between the magnets and make contact with each other. Students are taught that a magnetic field is responsible for the positive and negative forces exerted by a bar magnet.

A common science experiment, where a clear plastic chamber is fashioned, with a hollow tube running through the center of the chamber and metal filings are trapped within the inner and outer boundaries of the plastic chamber, can be used to demonstrate the presence of a magnetic field surrounding a bar magnet. Once a bar magnet is placed in the hollow tube in the center of the chamber, the metal filings respond by positioning themselves along the magnetic field lines generated by the bar magnet. See Figure 81. This rudimentary science experiment demonstrates that magnetic field lines exist in a three-dimensional pattern surrounding the bar magnet extending from the positive end of the magnet to the negative end of the magnet. Though rudimentary in nature, this simple plastic chamber demonstrates one of physics greatest unsolved puzzles.

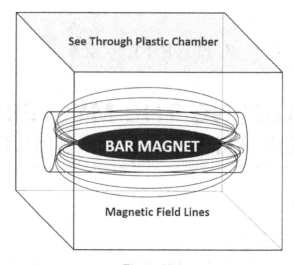

Figure 81

Illustration of a bar magnet's magnetic field lines using
a clear plastic chamber and metal filings

The luminous aether comprised of four poled quadsitrons offers an explanation as to how, in the proximity of a bar magnet, energy is trapped and continuously circulates in channels facilitated by the alignment of bands of quadsitrons to create the behavior of a fixed magnetic field surrounding a simple bar magnet. Similarly, aligning quadsitrons at the sub-sub atomic level, into channels within and surrounding the earth, to make it possible for energy to flow as a magnetic field, which protects life from the lethal radiation emitted by the sun. Channels of energy created by aligned quadsitrons, causes energy to surround and fall back toward earth, similar to the construct of bar magnet. The energy is trapped in loops and continuously circulates in and around a bar magnet, as well as the earth, at the speed of light within the invisible loops.

Quadsitrons, in the face of an organized structure as the material which comprises a bar magnet, can be organized into channels. These channels are capable of trapping energy. Tens of millions of channels may surround a bar magnet in the three-dimensional space surrounding the magnet. See Figure 82. Far more channels would comprise the magnetic field surrounding a celestial body such as the earth.

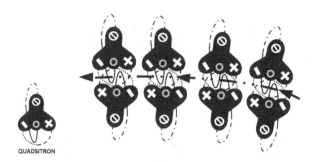

QUADSITRON

Figure 82

Quadsitrons organize their q-field lines to create channels.
The quadsitrons line up and the flow of energy is
channeled between such organized quadsitrons.

Three materials capable of becoming magnetized, which include iron, nickel and cobalt. Lodestone is a naturally occurring substance, which is magnetized and can attract small pieces of iron. The ancients learned of the existence of magnetism by the use of lodestone. At one time lodestone meant 'leading stone' due to pieces of lodestone were used to construct early compasses for navigation.[87] Lodestone is comprised of magnetite. Magnetite is one of the main ores of iron, an oxide of iron containing both bivalent and trivalent iron.

The crystal structure of a material, specifically one of the three magnetic materials of iron, nickel, and cobalt, are able to cause quadsitrons to organize into channels, not only within the boundaries of the material, but in the space immediately surrounding the magnetizing material. See Figure 83.

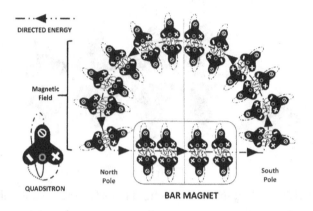

Figure 83

Quadsitrons organize into channels inside and
outside boundary of a bar magnet

Such magnetizing materials act as a sink for energy. Magnetized materials capture loose energy which flows in the universe, in proximity to the magnetizing material. Such loose or free energy becomes trapped in channels created by structuring the quadsitrons. This trapped energy continuously circulates in the channels until released by breakdown of the material or drawn out of circulation in the channels by another more magnetic and/or conductive material.

A bar magnet has numerous channels of energy circulating inside and outside of the boundaries of the material comprising the magnet. See Figure 84. In this example, there is no positive pole or negative pole to the bar magnet. The concept is that energy trapped by the materials comprising the magnet, continuously flows through the channels. For purposes of this illustration, the south pole is designated at the end of the magnet where energy flows out of the magnet and the north pole is considered the end of the magnet where energy flows into the magnet. The south pole of a bar magnet is likened to a geyser with energy flowing outward, and the north pole is likened to a drain, with the flow of energy into the boundaries of the magnet. The energy flowing through the magnet is of a very low frequency in the electromagnetic spectrum or simply a straight flow of energy with no measurable frequency. Since the flow of energy in the magnetic field surrounding a bar magnet has no frequency or may exhibit a frequency outside the bandwidth of visible light, the human eye cannot detect the energy circulating around the outer perimeter of the bar magnet.

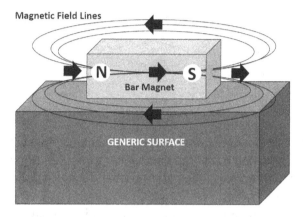

Figure 84

Bar magnet contains multiple channels of energy coursing inside and outside the boundaries of the material.

When two bar magnets are caused to approach each other, both have magnetic fields which extend beyond the physical boundaries of the material. See Figure 85.

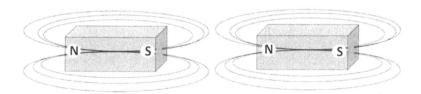

Figure 85

Two bar magnets, each projects a magnetic field beyond the physical boundaries of the magnetic material.

Two identical bar magnets will project magnetic fields in a similar fashion. The south pole of each bar magnet will act as a point of energy flowing out of the bar magnet. The flow of quanta of energy will follow individual channels in an elliptical pattern through and around the bar magnet. The channels exist in the three-dimensional space surrounding the bar magnet. The energy trapped in the channels formed by the quadsitrons will re-enter the bar magnet at the north pole. See Figure 86.

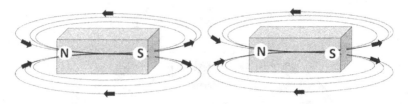

Figure 86

Two bar magnets, each projects a magnetic field with energy flowing out of the bar magnet from the south pole and re-entering the bar magnet at the north pole.

When two bar magnets rest on a surface and are separated by a space, gravity exerts a force to hold them down on the surface, friction created by gravity represents the resistance to movement from side to side. See Figure 87. It is observed that two bar magnets repel each other if like poles of each bar magnet are positioned in close proximity to each other. It is further observed that two bar magnets will attract each other if opposing poles of the two bar magnets are brought in close proximity to each other. If the distance between the two opposing poles of the two bar magnets is shortened sufficiently, the two bar magnets will exert enough force, without further assistance, to breach the remaining distance between the two bar magnets, collide, making contact with each other.

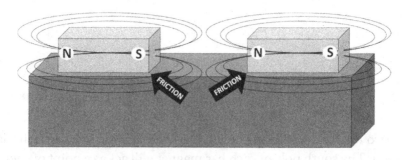

Figure 87

Two bar magnets separated by distance with friction restricting movement.

As the distance between two bar magnets shortens, the two bar magnets begin to share flow of energy in the magnetic field channels created by the alignment of quadsitrons which exist within and outside the boundaries of the two magnetic bars. See Figure 88.

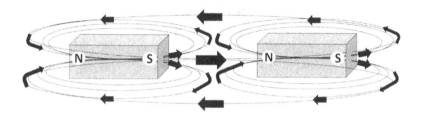

Figure 88

Two bar magnets in close proximity share energy flow in channels.

The act of sharing of energy by two bar magnets does not explain how physically the force of friction is overcome and two bar magnets are able to breach the distance to make contact. The common theory for the observed attraction between two bar magnets is the magnetic field of the opposing magnetic poles are the root cause, resulting in opposing magnetic fields cause inanimate magnetized objects to attract each other. Similarly, a magnetic field is thought to be the root cause by which a magnet is to attract a metal object, which can be magnetized, towards itself. Stating that a 'magnetic field' causes two weighted objects to physically more towards each other is an oversimplified explanation of the observed event and fails to describe the physical mechanism, at the subatomic level, required to move both bar magnets.

A deeper, scientific explanation must exist. It is hypothesized, that the convergence of energy trapped by the two bar magnets generates a funnel or a cyclone in the space between the south pole of one bar magnet and the north pole of the other bar magnet. This miniscule cyclone, present at the subatomic level of the quadsitron, redirects the quadsitrons present between the two bar magnets out of the way of the bar magnets, momentarily creating an empty space between the two bar magnets. See Figure 89. A gap in the fabric of the universe is not tolerated, except in extreme circumstances as the product of an immensely intense energy field, and therefore, the quadsitrons comprising the two bar magnets rapidly physically move in space to fill the gap located between the two bar magnets.

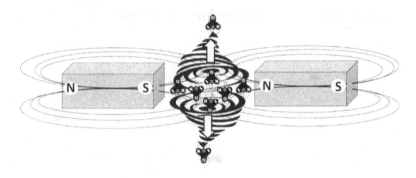

Figure 89

Cyclone is generated in sub² space between the two bar magnets.

Empty space is generally not tolerated in the universe, so the gap between the two bar magnets must be filled. The quadsitrons within the bar magnets are held in place due to the crystal structure of the material comprising the bar magnets, so the quadsitrons inside the magnets are not capable of moving to occupy the space, unless the bar magnets actually move. The bar magnets are pushed together by the compressive force of the three-dimensional ocean of a quadsitrons surrounding the bar magnets. The two bar magnets physically overcome the friction in order to seal the gap between the physical boundaries of the two bar magnets. See Figure 90.

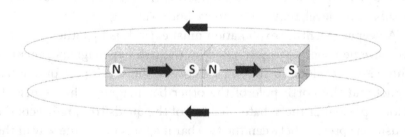

Figure 90

The space between the two bar magnets collapses, the two individual bar magnets become one bar magnet.

As mentioned above, and again seen in Figure 91, is illustrated two bar magnets with the south pole of one bar magnet facing the north pole of a second bar magnet. Given the configuration of a south pole of one bar magnet facing the north pole of a second bar magnet, the tendency is for the two bar magnets to exert an attractive force on each other.

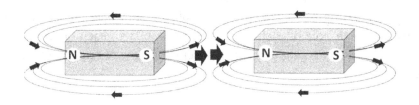

Figure 91

Two bar magnets, south pole faces north pole,
attractive forces are generated.

When two bar magnets are placed in proximity to each other, with the south pole of each magnet facing each other, repulses forces exist due to the flow of energy through both magnets, flowing out of each magnet through the south poles. See Figure 92. Both south poles act as a point of energy flowing out of the bar magnets. It would be similar to attempting to couple the ends of two garden hoses together while water was flowing out 0f the end of both hoses. Coupling together two geysers would be very difficult.

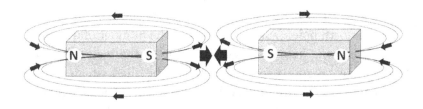

Figure 92

South pole of two bar magnets facing each other creates
repulsive force between the two bar magnets.

When two bar magnets are placed in proximity to each other, both magnets with the north pole of the magnet facing each bar magnet, repulses forces exist due to the flow of energy through both magnets and flowing into both magnets through the north pole of the magnets. See Figure 93. Both north poles act as a point of energy flowing into each of the bar magnets. The ends of the two bar magnets where the energy loops around to re-enter the bar magnet conflicts with each bar magnet, creating repulsive forces. Coupling is not possible by placing both the north pole of each magnet in proximity to each other due to both north poles acting a drain to accept the flow of energy into each respective magnet.

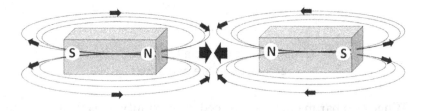

Figure 93

North pole of two bar magnets facing each other creates repulsive force between the two bar magnets.

In summary, when the south pole or the north pole of two bar magnets are placed in proximity to each other, the flow of energy is in the same direction, which causes a repulsive force between the two bar magnets. When a south pole of one bar magnet is placed in proximity to the north pole of another bar magnet, energy flows out of the south pole and into the north pole, causing a deficit in the number of quadsitrons, due to quadsitrons being maneuvered out of the way. The void of quadsitrons creates empty space, which must be filled, which causes the two bar magnets to be drawn together until the bar magnets breach the gap and make contact. The energy then flows through both bar magnets and creates the magnetic field around both bar magnets.

CHAPTER 25

SUBATOMIC ELECTRIC MOTOR

WIRE UTILIZED TO CONDUCT electricity is generally made of copper or gold. The standard explanation as to how a bar magnet can induce a current in a copper or gold wire is limited by a lack of definition of what a magnetic field actually represents at the sub atomic level. Many books have been written regarding how a magnetic field behaves, but what comprises a magnetic field and how it specifically interacts with copper at the sub atomic level remains to be defined.

The concept of an electric motor includes coiled wire, current flow and a magnet. Current induced in a coiled wire has the capability to move or rotate a magnet, which can be utilized as a motor to spin a shaft. See Figure 94. Alternately, energy can be transferred from a moving magnet to a coiled wire to create current in a generator.

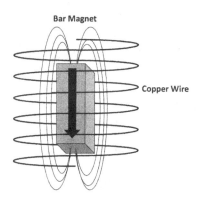

Figure 94

The basis of electromagnetic machinery: Energy transferred from a moving bar magnet to a coiled wire or vice versa.

Copper is an efficient conductor of electricity. Electricity is considered to be the flow of electrons through a medium, generally a conductor such as a copper wire, but given sufficient magnitude of current and voltage, electricity may be conducted through mediums such as water or air.

A length of copper wire, is comprised of individual copper atoms. Copper atoms generally should contain 29 protons, 29 neutrons and 29 electrons. The most common isotope of copper actually contains 29 protons, 34 neutrons and 29 electrons. Copper is therefore 29th element in the periodic table with an atomic weight reported as 63.546 amu. The electron orbitals about the nucleus of a copper atom are designated $1s^2$, $2s^2$, $2p^6$, $3s^2$, $3p^6$, $3d^{10}$, $4s^1$. A valance electron is an electron in outer 4s orbital that has the propensity to enter into a covalent bond with another atom.

A magnetic field represents a flow of quanta of energy trapped in loops created by channels of quadsitrons existing inside and outside a magnetic material. See Figure 95. There are numerous loops of energy surrounding a bar magnet. This trapped energy can be coaxed to run through a copper wire as the bar magnet is passed perpendicular to the coils of wire. The bar magnet is a perpetual reservoir of energy. Once energy is released from the bar magnet to the coiled wire, the magnet absorbs and traps an equivalent of free energy from the surrounding environment.

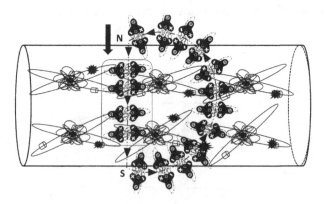

Figure 95

Bar magnet surrounded by channels of energy flow, passes ninety degrees to a copper wire. Flows of quanta of energy trapped in loops about a magnet, can pass to copper atoms in a coiled copper wire to generate a current in the copper wire.

As a magnetic field associated with a bar magnet moves across a copper wire at ninety degrees to the length of the copper wire, the abundant quanta of energy flowing through the loops of the magnetic field surrounding the bar magnet force quanta of energy into the valence orbitals of the copper atoms. The added quanta of energy introduced into the copper wire from the moving magnet, results in electrons flowing through the copper wire represented as quanta of energy, acting as electrons, traveling through the valence orbitals of the copper atoms which is measured as current. See Figure 96.

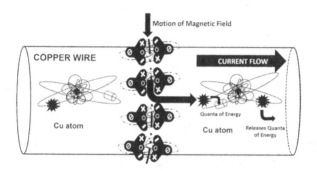

Figure 96

Quanta of energy being pushed through the copper atoms of a copper wire.

The movement of a bar magnet through coils of a copper wire, transfers quanta of energy from the many channels of energy which surround the bar magnet to the copper atoms comprising the coiled wire to create current in the copper wire. See Figure 97. The copper wire must be a part of an electric circuit. A simplified electric circuit usually includes wire and a load. A load is some form of use of electricity such as a light or a motor. Electric circuits can be built very complex with resistors, capacitors, and various other electronic components.

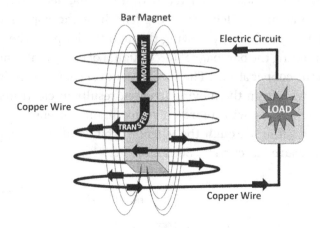

Figure 97

Movement of a bar magnet through coils of copper wire creates current.

A bar magnet moving through coils of copper wire creates a current. Though the magnet gives up quanta of energy to the coiled wire, the bar magnet never truly loses its magnetic field. The bar magnet acts as a sink for energy in the universe. Once a magnet loses quanta of energy to the coiled wire, the magnet traps other stray energy passing through the luminous aether in close proximity to the bar magnet.

Since a bar magnet acts as a perpetual sink for energy, when a bar magnet is at rest, the bar magnet is not able to absorb more energy. When a bar magnet at rest is exposed to a magnetic field flowing in a coiled wire, the quanta of energy comprising the magnetic field cannot be absorbed by the bar magnet, therefore this creates resistance between the energy in the magnetic field surrounding the coiled wire and the energy captivated by the bar magnet. If sufficient current courses through a copper wire, which is coiled around a stationary bar magnet, enough force may be exerted on the bar magnet to cause movement of the bar magnet from one location to another location in space. The energy already stored in the bar magnet, which is at capacity, creates a resistance, which can be acted upon by the force created by the energy in the magnetic field surrounding the wire coiled around the bar magnet. Again, this is the fundamental principle of utilized in the design of electric motors.

CHAPTER 26

EARTH AS A BAR MAGNET

THE EARTH WE LIVE on is comprised of numerous elements. Much of the mass comprising the planet is ferrous material. Due to the large amount of ferrous material comprising the material substance of the Earth, the planet as a whole, functions as a gigantic cosmic bar magnet. See Figure 98.

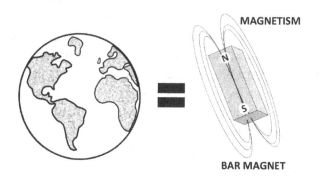

Figure 98

Earth as a bar magnet.

Like a simple bar magnet that one can hold in their hand, the Earth has a north and a south pole. It has been said that energy flows out of the Earth's south pole and into the Earth's north pole. Due to this nomenclature, this is why the north and south pole of the bar magnets described previously were designated in the manner presented. See Figure 99.

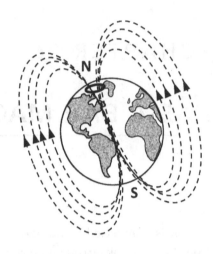

Figure 99

The Earth with a north pole and a south pole.

The Earth, like a bar magnet, organizes quadsitrons into channels to trap energy. The massive ferrous materials comprising the Earth cause quadsitrons to line up inside the planet and to surround the planet in a grand scale. The energy circulates through the core of the planet and around the planet within the channels created by the aligned quadsitrons. See Figure 100. The Earth traps free energy passing through the surrounding universe on a colossal scale. Once the energy quanta are trapped, they continuously and endlessly circulate around the surface of the planet, then looping through the core of the planet. When the Earth's bar magnet is at capacity, by having trapped the maximum energy, the magnetic field cannot be increased further and additional energy emitted by the sun or passing through the solar system from an outside source, is diverted away from the Earth. This diversion of excess radiant energy protects the organic life inhabiting the planet. The bar magnet effect of the planet makes human life and the vast ecosystem which supports our existence, possible.

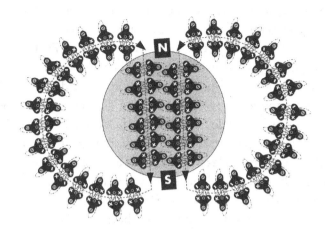

Figure 100

Energy circulates through the Earth and around the Earth
in channels created by alignment of quadsitrons.

In a simple bar magnet one can hold in their hand, the numerous circulating quanta of energy trapped in channels in and around the bar magnet are invisible, but detectable as seen by the resistance expressed by trying to move the like ends of two bar magnets into close proximity to each other. Resistance is detected when the north pole of two bar magnets are brought together. Resistance is also detected when the south pole of two bar magnets are brought together. The energy circulating in and about the Earth comprises the magnetic field that protects the Earth from lethal radiation coursing through space. See Figure 101.

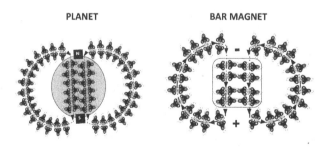

Figure 101

A bar magnet exhibits a magnetic field, the Earth
exhibits a similar magnetic field.

Humans can detect the presence of the magnetic field surrounding the Earth by viewing the Aurora Borealis. The Northern Lights or Aurora Borealis also seen at the southern pole of the Earth, referred to as Aurora Australis, is the visual display of charged particles emanating from the sun striking the protective magnetic field surrounding the Earth.[88] The magnetic field about the Earth deflects most of the highly charged, dangerous particles emitted by the sun. At the northern and southern poles of the planet, some of the charged particles from the sun collide with gaseous particles in the Earth's atmosphere and emit energy in the visible light spectrum. The Aurora Borealis typically occurs 90 km to 150 km miles above the Earth's surface and is seen as bright dancing lights illuminating a night sky in the northern latitudes.

CHAPTER 27

EARTH AS A PUMP

IT IS NOT ENOUGH to say that the Earth is a large bar magnet. The current concept of a magnetic field surrounding the Earth technically does not include the concept of gravity. A magnetic field and a gravitational field have been classified as two different physical phenomena. But what if separate classification of these two forms of energy was not accurate. Possibly the sole difference between a magnetic field and a gravitational field is the density of the energy used to create the two phenomena. Possibly the energy comprising gravity is denser than the energy comprising a magnetic field. Possibly, the difference between the force of gravity and the force of magnetism, is that gravity is dense enough that it is capable of physically moving quadsitrons, where magnetism may be able to align quadsitrons to create paths of energy flow, but not sufficient enough in force to generate a flow of quadsitrons.

Energy can be created in a wave pattern, as described by the electromagnetic spectrum. A limited portion of the electromagnetic spectrum is visible to the human eye referred to as visible light. Energy may be created and may move as straight-line flows of energy, which can include magnetism and gravity. Straight line flows of energy or energy with very low frequencies outside the visible light spectrum, and therefore exist about us as physical properties, but are invisible to the human eye.

If gravity is related to the density of energy such that the density of flow is capable of moving quadsitrons in space, then gravity may be likened to energy in a body of water. If water molecules, H_2O, are likened to quadsitrons, and currents that run in a body of water are likened to gravity, then all sorts of physical possibilities exist and can be explained with the behavior of water as a point of reference.

Gravity has been thought of being an unseen force pulling an object from the core of a heavenly body. See Figure 102. An object on the Earth's surface or in the lower atmosphere is thought to drop from a higher elevation to a lower elevation due to the force of gravity pulling the object toward the core of the planet. But what if the core of the planet was instead a pump, and with a dense enough flow of energy, capable of exerting forces on quadsitrons, to effect a flow of quadsitrons through the planet. That is, if gravity was instead a flow of quadsitrons from outer space, through the planet, causing objects in close proximity to the planet and on the surface of the planet to be pushed toward the center of the planet this would create the mechanism for what is recognized and known as the force of gravity. The flow of quadsitrons from space would push down against every proton and neutron comprising every atom of an object, causing the object to preferentially desire to follow the flow of quadsitrons toward the core of the planet. All objects comprising planet earth, residing on the surface and in close proximity to the planet would experience the force generated by the flow of quadsitrons streaming toward the core of the planet. Since the only effect which occurs is located on the surface of protons and neutrons, the flow of quadsitrons toward the center of the planet fails to create any detectable wake.

Figure 102

The phenomenon of gravity is created by a dense enough force causing quadsitrons to stream down out of the heaves above and drive all objects toward the center of the planet.

It has been taught that the core of the Earth is molten. We have been taught that the core of the planet is comprised of metals, heated to the point they act as fluids. This molten core at the center of the Earth is continuously churning. Given the pressures and continuous movement of the core at the center of the planet, it acts as a pump directing the flow of quadsitrons. See Figure 103.

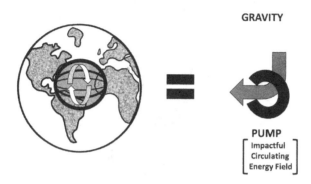

Figure 103

The cyclic compression and expulsion action of Earth's molten core acts as a pump attracting quadsitrons from space toward the center of the planet and then ejecting these quadsitrons back out into space.

Like any pump, the Earth's molten core draws in the material it is pumping and ejects the same material with force. In the case of the Earth's molten core, the molten core draws in quadsitrons from outer space and then ejects the quadsitrons back out into space. See Figure 104.

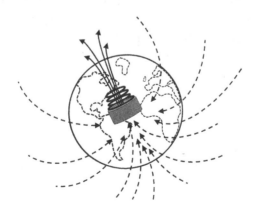

Figure 104

Earth's molten core draws quadsitrons in from outer space from all directions, then ejects the same quadsitrons back out into space.

If the molten core of the planet were stationary, one would expect a large hole to be present in the Earth's surface where the quadsitrons would be continuously ejected. But the molten core at the center of the Earth, continuously moves; endlessly rotating and gyrating on its axis. See Figure 105.

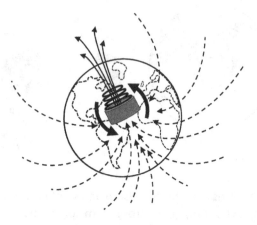

Figure 105

Molten core of the Earth continuously tumbles in place which causes the direction of output to be reoriented with every pump of the molten core.

By the fact that the Earth's molten core continuously re-aligns itself, this results in the ejected quantity of quadsitrons being directed back out into space at differing locations spread out over the entire surface of the planet. The effects of gravity are relatively uniform across the planet's surface, because all of the quadsitrons in space are continuously attempting to move toward the Earth's active core in order to fill the empty gap left by the core's action of ejecting quadsitrons out into space. See Figure 106.

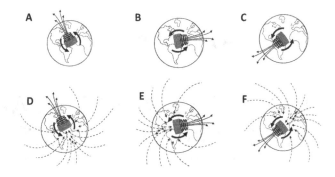

Figure 106

The pumping action of the Earth's molten core changes direction with each compressive output of the pump. The letters A, B, C, D, E, and F demonstrate differing directions of the quadsitrons outflow of the pumping action of the molten core with each pumping action generated by the molten core. The net result as demonstrated in illustrations D, E, and F is that the pump continuously draws quadsitrons in toward the core of the earth from all directions, but each pumping action ejecting these same quadsitrons is delivered in a different direction.

No specific hole is detectable in the Earth's crust due to the fact that the pressure of the pump and the ejection of quadsitrons changes direction along the spherical surface of the planet with every discharge of quadsitrons by the molten core's pumping action. Thus, the direction of the output of the pump continuously changes, but paradoxically the input into the pump from surrounding space maintains a constant rate with the gap created by the pumping action, filled with quadsitrons from every direction of three-dimensional space. The fact that quadsitrons are perpetually drawn toward the core of the planet from all directions is the physical basis of the phenomenon recognized as the force of 'gravity'.

CHAPTER 28

WHY DOES THE APPLE REALLY FALL FROM THE TREE?

SIR ISAAC NEWTON DESCRIBED the reason for the apple falling from a tree to the ground was due to gravity. *Memoirs of Sir Isaac Newton's Life* composed by William Stukeley, one of Newton's first biographers, published in 1752.[89] Newton told the apple story to Stukeley, who relayed it as such:

"After dinner, the weather being warm, we went into the garden and drank tea, under the shade of some apple trees...he told me, he was just in the same situation, as when formerly, the notion of gravitation came into his mind. It was occasion'd by the fall of an apple, as he sat in contemplative mood. Why should that apple always descend perpendicularly to the ground, thought he to himself..."

So, gravity is the culprit, causing an apple to fall to the ground. See Figure 107. Gravity has long been thought of as some energy force pulling objects on the Earth surface toward the center of the planet.

Figure 107

Gravity is thought to pull on an apple and when the bond between the apple and the tree are broken, the apple falls to the ground due to the unseen force of gravity.

As discussed in the previous chapter, what if gravity was actually a dense enough energy to cause quadsitrons to move. As the Earth's molten core pumps quadsitrons out from the center of the planet into the surrounding space, quadsitrons from outer space race from all directions surrounding the sphere-shaped planet to fill the gap created by the pumping action of the molten core of the planet. The movement of quadsitrons toward the center of the planet strike the protons and neutrons of every atom comprising any material entity, forcing all objects with mass comprising the planet or resting on the surface or located in close proximity of the planet, in the direction of the core of the planet.

In the context of an apple, the quadsitrons streaming down from outer space strike the protons and neutrons of every atom comprising the apple, forcing the apple toward the surface of the planet, which is the direction of the flow of quadsitrons from outer space. See Figure 108. This is similar to an object caught in a moving river. The current of the waters of the river will force the object in the direction the river is flowing, unless the object is able to generate a force equal to or greater than the force of the current.

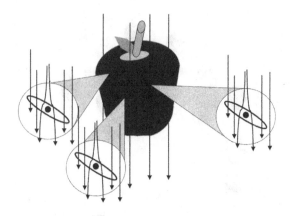

Figure 108

The flow of quadsitrons act as a force pushing the protons and
neutrons of each atom down towards the center of the planet.

As a whole, the quadsitrons flowing from outer space down toward
the core of the planet act on the entire apple tree. See Figure 109. The
reason the apple falls to the ground is that the stem, holding the apple
to the tree branch, becomes dry, fragile, eventually breaking. Upon
separation of the apple from the tree, the apple is forced to the ground
by the constant, relentless stream of quadsitrons flowing from outer
space in the direction of the center of the planet.

Figure 109

Quadsitrons flowing from outer space toward the core
of the planet act on the entire apple tree.

Similarly, in the human body, as any object existing on the planet's surface or in the immediate atmosphere, the protons and neutrons comprising all of the atoms are acted upon by the flow of quadsitrons moving from outer space toward the center of planet Earth. See Figure 110.

Figure 110

The flow of quadsitrons act on the protons and neutrons at the center of all of that atoms comprising the human body creating the effect of gravity.

As with the example of the apple, the core of the planet acts as a pump ejecting quadsitrons out into space, thus creating a deficiency of quadsitrons at the core of the planet. See Figure 111. The deficiency created at the center of the planet by each pumping acting of the molten core, must be filled with quadsitrons, therefore, quadsitrons rush in from outer space to fill the void at the center of the planet. This constant flow of quadsitrons, from outer space toward the core of the planet, creates the observed gravitational effect. All objects resting on the planet, are residing on the planet at the lowest altitude possible given the surface of the planet, due to the incessant stream of quadsitrons raining down from the heavens above, passing through each molecule and unremittingly urging all matter in the direction of the core of the planet from all directions, creating a sphere of matter. It can be thought of as standing under a perpetual gravity waterfall pouring down from the sky above.

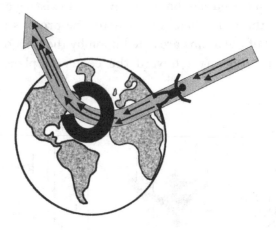

Figure 111

The pumping action of the molten core of the Earth generates
a constant flow of quadsitrons directed toward the center of
the planet which is interpreted as a gravitational effect.

Potential energy is thus, simply, based on an object with mass, length of potential travel from height A to height B with the scope of constant flow of quadsitrons. The constant migration of quadsitrons toward the center of the planet creates a standard acceleration of gravity in free fall near the surface of the earth, measured as 9.807 m/s.[38]

The Earth, as a celestial body, acts both as a bar magnet exerting a magnetic force into space and as a pump, generating a gravitational effect drawing quadsitrons from out in space in toward the core of the planet. See Figure 112.

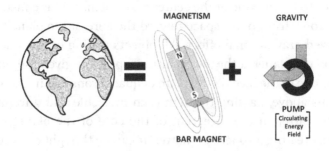

Figure 112

The Earth acts as a bar magnet and as a pump.

It is the dual action of the Earth acting as a bar magnet and as a pump which makes life possible. The magnetic field created by the Earth protects organic life from the lethal radiation emitted from the sun. The pump action of the core stabilizes the atoms comprising the planet, creating order and generating the observed gravitational effect, which creates an environment conducive to life.

Returning to the interpretation of the Michelson and Morley experiment of 1887, the concept is that the quadsitrons are so small, that most pass through the human body, without interruption. Thus, the majority of the medium for energy transfer, passes through the human body unimpeded. Low levels of radiant energy, such as sunlight exposure at the surface of the planet, tends to be absorbed by the skin, and superficial layers of tissue just under the skin. Large doses of high-frequency energy, such as x-rays, when directed at the human body would only be partially absorbed by only the significantly dense tissues, such as bone; the remainder of the energy passing through the tissues of the body. Such high frequency x-ray energy passing through the human body is routinely utilized by medical professionals to generate radiographs to investigate for evidence of certain diseases.

Some of the energy of high dose radiation would be absorbed by the electrons of the atoms comprising the body's tissues. Damage to the tissues of the body is generally dose dependent. Short exposures to high energy radiation might disrupt the fragile bonds of the atoms comprising the DNA of cells, leading to the development of certain cancers such as leukemia. If the dose of radiation were significant enough, actual immediate tissue damage would occur due to electrons incurring too high of a dose of radiation. When molecules are vibrating too vigorously and covalent electrons are overloaded with energy, molecular bonds are broken, which can destroy the molecules which comprise cells. As the destruction of molecular structures becomes increasingly widespread, this results in tissue breakdown, which can lead to organ failure. Prolonged exposure to high doses of radiation can lead to the superficial skin being noticeably burned, liver and kidney failure, fibrosis of the lungs, shut down of the bone marrow, damage to the intestines and muscle structures. The fact that energy such as x-rays can pass through the human body and create radiographic pictures of the biologic structures, spawns the concept that the fabric of the universe also passes through the human body.

CHAPTER 29

SUN AS A TWISTING BAR MAGNET & PUMP

THE SUN ACTS AS a bar magnet and a pump, similar to the Earth, but with a twist. The sun's core is so active the center of the star continuously changes in 3D space, generating fluctuating magnetic and gravity fields. See Figure 113.

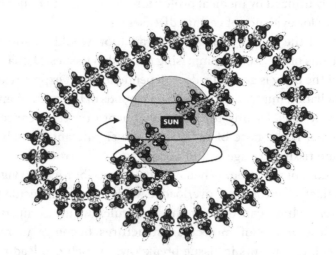

Figure 113

The Sun exerts a fluctuating magnetic and gravity fields.

The position of the planets orbiting the Sun are present in their locations due to the balance between the mass of the planet, forward direction of the movement of the planet, flow of quadsitrons in toward the Sun and the flow of quadsitrons being ejected by the Sun. See Figure 114.

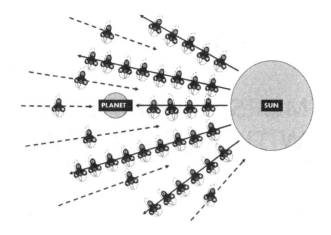

Figure 114

Planetary positions are dictated by the balance of mass of the
plant and the flow of quadsitrons in and out of the Sun.

Over time the planets have reached a point of balance given their
mass, orbital velocity, original position at the conception of the solar
system, and the flow of quadsitrons in and out of the Sun at the center
of the solar system.

CHAPTER 30

MARS, BAR MAGNET WITH A DEAD CORE

MARS IS THE FOURTH planet in the solar system. Mars is often referred to as the red planet. The planet is arid, the surface void of appreciable water except for possibly at the poles or hidden in underground reserves. The majority of Mar's atmosphere as gathered and reported by the Curiosity Rover, includes 95.97% carbon dioxide, 1.93% argon, 1.89% nitrogen and 0.146% oxygen.[90] By comparison, Earth's atmosphere, dry air by volume, is comprised of 78% nitrogen, 20.9% oxygen, argon 0.9%, 0.04%. carbon dioxide and a variety of other trace gases.[91] Mars' atmosphere is inhospitable to humans. Massive dust storms often cloud and consume much of the surface of the planet for months at a time.

Mars is thought to have an inactive core. Without a molten core there is no volcanic activity; there is also no internal pump to create a high gravitational energy flow. The gravitational acceleration on Mars is 3.72 m/s^2, versus the gravitational acceleration on the Earth which is 9.807 m/s^2.[92] The gravitational force on Mars is thought to be 38% that of Earth's gravity. Thus, a 70 kg man on Earth would weigh 26.6 kg on Mars. The size of Mars is 53% that of the Earth.

The mass of the Earth is 5.972 x 10^{24} kilograms, where the mass of Mars is 6.42 x 10^{23} kg; thus, Mars is considered to have a mass approximately one tenth that of the Earth. But mass is an entity which is measured in relation to the force of gravity, and the force of gravity differs in relation Mars versus the Earth. A 70 kg man with a mass of 70 kg on the Earth's surface, but if deployed as an astronaut in space, this same 70 kg man would be weightless. The standard for comparing planets would be to compare both the number and type of atoms which comprise each planet. The number of atoms comprising the earth has been calculated to be 1.33 x 10^{50}. [93] The exact number of atoms and specific types of atoms comprising the Earth versus Mars is beyond the scope of current technology. Neither quantifying Mars to the Earth by mass or by number/type of atoms provides an accurate measure of comparison for the purposes of this subject matter.

The equatorial radii of the Earth is reported to be 6,378 km, where the equatorial radii of Mars is 3,396 km.[94] Neither planet is exactly a sphere, but more of an oblate spheroid due to the flattening which has occurred at the poles due to the constant rotation of both planets. Still the volume measurement for a sphere is volume = 4/3 x π x r^3. The volume of the Earth is reported to be 108.321 x 10^{10} km^3. The volume of the Mars is reported to be 16.318 x 10^{10} km^3. Thus, though the diameter of Mars is 53% of the diameter of the Earth, the volume of Mars is only 15% of the volume of the Earth.

Based on the measure of volume, gravity on Mars should be 10% of that of the Earth. Based on the physical size of volume, the gravity of Mars should be 15% of the Earth. As mentioned above, the gravitational force of Mars is 38% of that of the Earth. The weak gravity that Mars exerts is possibly based on the composite of ferrous materials comprising the planet, which causes the red planet to act as a cosmic bar magnet. See Figure 115. The percentage of magnetic materials comprising the overall material of a planet, most likely result in the quantity of a planet's passive gravitational field.

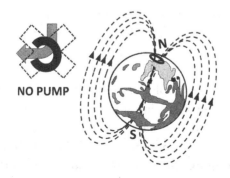

NO PUMP

Figure 115

Mars acts as a large bar magnet, but due to a lack of a molten core there is weak gravitational effect.

Having no active core, Mars has no active pumping system present at the center of the planet. Without an active molten core, Mars is not able to actively draw quadsitrons in from the surrounding space and then eject these same quadsitrons back out into space. Without an active molten core to act as a celestial pump, Mars is not able to generate a comparable gravitational force to that of the Earth. The weight of an object on Mars is solely based on the ferrous material composition of the planet creating a passive gravity force.

The reason the gravity of Mars is 38% that of the Earth, rather than comparable to that of the Earth, is that Mars exhibits only a passive gravitational force, rather than the Earth's combined passive and dynamic gravitational forces. It is likely that relative size of a planet dictates the capacity of the active core to generate a gravitational force; that is, given two planets with a molten core, made of the same materials, the larger the planet the greater the composite gravity.

It should be noted, that the capacity of a molten core to create a gravitational effect, which is capable of generating a dynamic gravitational force for a celestial body, may not have any relation to the physical size or mass of the celestial body. The efficiency of the molten core to pump quadsitrons may be different for differing celestial bodies. This, in part, may be related to differing materials comprising differing celestial bodies. Thus, a smaller celestial body with a molten core exhibiting a high pumping efficiency, may indeed generate a stronger gravitational force on the surface of the planet than a larger celestial body with a molten core which is less capable at attracting and pumping quadsitrons.

Thus, if at one time Mars had a more efficient molten core than that of the Earth, though Mars is smaller in size, the red planet may have had a gravity similar to that of the Earth.

Long ago Mars may have supported life, but given the lack of an active core Mars is a dead planet and unfortunately the atmospheric conditions make the red planet a rather inhospitable choice for human colonization; still it may be our only choice of extraterrestrial expansion if the need to venture beyond the safe haven of the Earth were to press humans into interplanetary exploration on a grand scale.

The Earth's Moon is of similar construct. The Moon has a dead core, if it ever had a core at all. Therefore, the 1/6th gravity the Moon has in comparison to the Earth is based on the Moon, as a celestial body, acting as a huge bar magnet. The metals comprising the Moon create the weak gravitational field which passively surrounds Earth's lone satellite.

CHAPTER 31

OUR WONDERFUL WORLD OF COLOR

COLOR IS SIMPLY THE product of what the sensors comprising the human eye are capable of detecting and what the human brain is capable of interpreting. When we look at an image of ourselves in a mirror, we typically see a variety of shapes and colors. See Figure 116. Photoreceptor cells are present in the posterior portion of the human eye, referred to as the retina. The light sensitive tissues comprising the retina are made of cones and rods. Cones are photoreceptor cells, which function in daytime, and brightly lit environments to provide high acuity and color vision. Rods are nighttime photoreceptor cells, which provide black and white vision during low light environments.

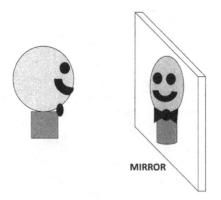

Figure 116
A person viewing oneself in a mirror.

When looking at the reflection of oneself in a quality mirror, colors and imagery are cast back in a manner that tends to coincide with the original physical form. See Figure 117.

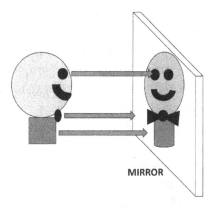

Figure 117

Reflection of oneself in a quality mirror, imagery cast back tends to coincide with the original physical form.

The human visual sensory system is comprised of the eyeball, the retina, the optic nerve, and the occipital lobes of the brain. See Figure 118.

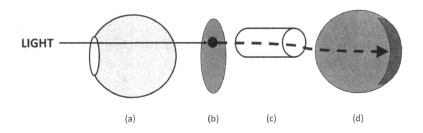

Figure 118

Visual sensory system is comprised of (a) eyeball, (b) retina, (c) optic nerve, and (d) occipital lobes of the brain, which are located at the back of the head.

The eyeball is a spherical organ. The construct of the eyeball is used to optimally direct light gathered by the pupils, to the retina lining the posterior portion of the eyeball. Photoreceptive cells embedded in the retina collect the incoming light and convert the light to electrical signals. The electrical signal is carried by the optic nerve to the back portion of the brain. The posterior or back portion of the brain is comprised of the occipital boles of the brain, one on the right and one on the left. The occipital lobes download the electrical signals input from the retina of the eyes and interprets the imagery of light signals. Images constructed from the data provided by the retina are compared to known images previously learned and stored in the brain long-term memory. The comparison of the active sight as seen by the eyes, to known image files, cross-referenced to language and learned data regarding background associations to known image files, provides understanding of visual imagery in real-time.

In the case of a mirror, quanta of white light initially strike an object which will appear in the mirror, such as in this example a human face. The atoms of the human face vibrate, causing quanta of energy to ripple out of each differing atom comprising the face, with different frequencies, resulting in a wide range of colors being emitted by the human face. The light of differing frequencies being emitted by the human face strike the surface of the mirror. The arrangement of the atoms comprising the mirror have the quality of absorbing the incoming light frequency, which creates a vibration phenomenon such that an identical outbound light frequency is emitted from the surface of the mirror. The higher the quality (lack of imperfections) of the mirror, provides a closer match of the inbound frequencies of light to the outbound frequencies of light, resulting in the mirror casting a greater resolution of image of the object positioned in front of the mirror.

CHAPTER 32

FASCINATING BEHAVIOR OF A RAINBOW SOLVED

THE BEHAVIOR OF LIGHT has captivated the imagination and curiosity of many students of science over the last several centuries. Sunlight strikes a rain shower falling from the sky, and viewed from the proper angle, the falling rain radiates a dazzling rainbow in the direction of the viewer. See Figure 119.

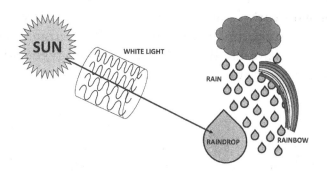

Figure 119

Sunlight striking water droplets creates a rainbow.

Amongst any rain shower where a rainbow is evident to a viewer, sunlight strikes individual water droplets, which creates the colors of the visible light spectrum. See Figure 120.

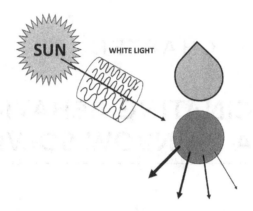

Figure 120

Sunlight strikes a water droplet.

The human eye detects a rainbow, if the point of observation is at an angle of 42 degrees to the original the direction of the source of sunlight.[95] The water droplets in a rain shower act similar to a prism. A shower of raindrops spread the colors of visible light out to create the rainbow effect. See Figure 121. How this actually occurs represents one of the most curious mysteries of physical science. Visible light is known to be a subpart of the electromagnetic spectrum. Therefore, light travels in a wave, like the remainder of the energy comprising the electromagnetic spectrum.

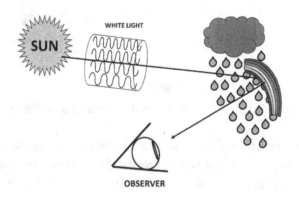

Figure 121

Reflection of light from raindrops, which acts like a prism, is referred to as a rainbow.

Reflection, light being casted back by an object, and refraction, the change in direction of a wave of light as the light passes through a medium, are physical phenomenon easily detected by the human eye. Current physics takes the stand that there is no luminous aether. One can say light bounces off the surface of an object, therefore creating the phenomenon of reflection, and in a macro sense where human senses tell us that objects are solid including atoms, it is easy to accept such a concept provides a sufficient explanation. But breaking visible light down to the quanta level and atoms down to the subatomic level, the theory that light bounces off objects begins to lose validity as one tries to apply atomic physics to explain the phenomenon of reflection. With no luminous aether underpinning the current theory, the mechanics of the science just does not make sense as a means of explaining what one observes.

For example, sitting in the audience, awaiting an orchestra recital, numerous musicians sporting a broad array of instruments occupy the stage. Similar types of instruments are usually organized together in groups. Violins usually sit with other violins, as do trumpets with trumpets and trombones with trombones. See Figure 122.

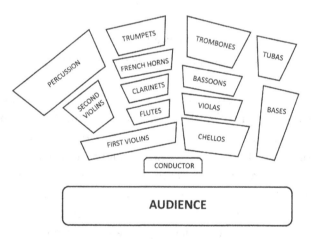

Figure 122

Various instruments comprising a typical orchestra.

Each type of instrument generates its own signature sounds. Listening to an orchestra play, the construct of the outer ear captures much of the sound generated by the individual musicians and funnels the sound to the inner ear canal. See Figure 123. The sounds resonate the tympanic membrane of the ear, transferring the sound energy to the cochlea of the ear. The spiral shaped cochlea decodes the sounds, converting the sounds to electrical impulses. The brain detects the collective of sound and in the case of an orchestra playing, interprets the sounds as music If the orchestra is playing well, the stream of sound generated by the orchestra should be pleasant to listen to by the members of the audience.

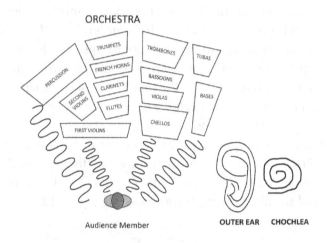

Figure 123

An orchestra playing generates sound in various differing frequencies.

The various waves created by the orchestra reach the human ears residing in the audience by the sound stimulating the air molecules, which act as the transfer medium. In a vacuum, there would be no sound. In an environment where there was no transfer medium for sound energy to be conducted from one location to another location, sound would not be able transfer from one point in space to another point in space. Sound is a form of wave energy. Light is a form of wave energy. It is reasonable to invoke that both wave energies require a transfer medium.

White light is generally comprised of much of the various frequencies of visible light. See Figure 124. The human eye takes in white light and the human brain, given the proper level of learning and experience, interprets the collective visual spectrum of light as the color 'white'. The absence of visible light is considered to be the color 'black'.

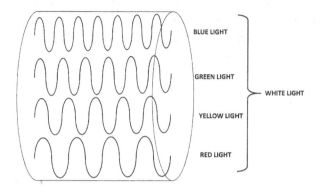

BLUE LIGHT

GREEN LIGHT

WHITE LIGHT

YELLOW LIGHT

RED LIGHT

Figure 124

White light is comprised of the majority of the
frequencies of the visible light spectrum.

The act of reflection is most commonly associated with a mirror. One looks into a mirror and sees a rather identical image of themselves staring back at them. See Figure 125. In this case, light is cast back from the mirror in almost the identical form that visible light struck the mirror surface. A ray of white light reflects back to the viewer as a ray of white light. Specific colors reflect back from the mirror's surface as the same color.

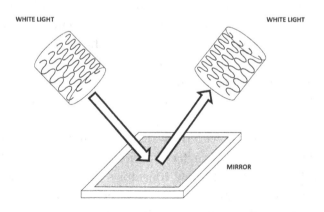

WHITE LIGHT   WHITE LIGHT

MIRROR

Figure 125

Light striking a mirror surface is reflected with the
same frequencies of energy as the initial light.

175

The supposed act of reflection is common in everyday life. All color is in essence created by reflection. White light strikes an object and the color or colors inherent to the object are cast back from the object to be detected by the human eye. See Figure 126.

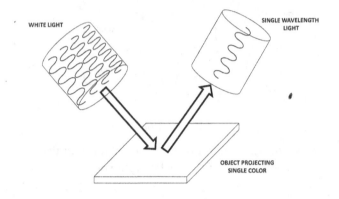

WHITE LIGHT

SINGLE WAVELENGTH LIGHT

OBJECT PROJECTING SINGLE COLOR

Figure 126

White light strikes the surface of an object and the frequency of color inherent to the object is cast back for the human eye to detect as color.

The current theory of physics suggests that reflection simply happens. The concept is that surfaces of objects are solid and white light is cast back from the surface of a solid object as the color inherent to the object being struck by light. An analogy would be throwing a tennis ball at the side of a brick building. Once the tennis ball strikes the solid surface of the solid wall of the building, the lighter mobile density of the tennis ball encounters the greater fixed density of the building's wall and the tennis ball is forced to retreat. The tennis ball's resultant trajectory following collision with the fixed wall is that of reflecting off the building's wall. The resultant trajectory tends to be dependent upon the initial angle of impact with the wall, the inflated pressure of the ball, integrity of the wall and the surface architecture of the wall at the point of impact. The tennis ball's new trajectory will generally be in some direction back toward the thrower. We often rationalize the behavior of visible light regarding reflection, with the oversimplified explanation that light bounces off of the surface of an object. See Figure 127. That is, reflection is caused by light bouncing off the electron orbitals of atoms comprising an object.

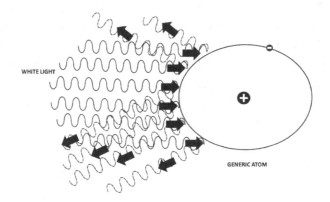

Figure 127

Light seemingly being reflected or bounced off atoms.

We are taught atoms represent solid structures. Again, this is a matter of illusion created by human senses. The construct of an atom is actually not solid, except for possibly the center of an atom where the protons and neutrons are congregated. It has been taught that the electrons orbiting the center of an atom move at the speed of light and therefore are in effect, an electron is in all places in an orbital pathway, all at the same time; thus, creating not the illusion, but the functionality of a solid. This concept might be feasible for the phenomena of touch and pressure, but to a quanta of light striking an atom, it is highly unlikely that light, traveling at a speed equal to an electron, and inbound to an atom, is bouncing off an electron orbital and being cast back out of the boundaries of an atom. Similarly, it is highly unlikely that an inbound quanta of light energy strikes the nucleus of an atom and is cast back out of the atom. The nucleus of an atom comprises such a small volume compared to the overall volume of the atom and a quanta of energy is so small, it would seem to be a rare event for a quanta to actually strike the center of an atom. Additionally, given the center of the atom is comprised of protons and neutrons, the configuration of the striking surface of the center of the atom would be dependent upon the angle the quanta would encounter the center of the atom, resulting in variance in the energy reflected by a material. See Figure 128. The structure of an atom as dictated by the current Electron Shell Model must include a nucleus which is rather oddly shaped. Frequently, across the array of elements comprising the

Periodic table, an atomic nucleus is not symmetrical in three-dimensional space due to the arrangements of protons and neutrons at the center of the atom. Light striking an atom from differing angles should result in differing responses; that is the light signature of an object should change if the position of the light source changes.

Quanta of light striking an atom resulting in reflection requires a more in-depth explanation.

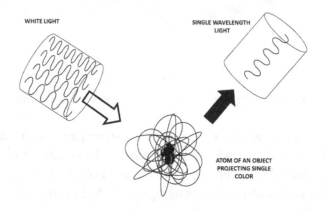

Figure 128

Quanta of light striking an atom resulting in reflection
is in need of a more in-depth explanation.

Further, if the theory that quanta of energy comprising a beam of white light were to bounce off an object to create the color of the object, then the theory of atoms absorbing quanta of light begins to unravel. It is known that quanta of light and other frequencies of the electromagnetic spectrum, enter the perimeter of an atom and result in electrons migrating to more distant orbitals from the center of the atom. Absorption of energy from the electromagnetic spectrum by an object sometimes results in increased temperature of the target mass. It would seem contradictory to accept that quanta of energy can be absorbed by electrons orbiting an atom in one instance, but bounce off the electrons of atoms in another instance. If this concept were representative of what is observed, the subatomic function of a mirror becomes somewhat blurred. Given it is likely a portion of the energy striking a mirror would be absorbed by the atoms comprising the mirror, it would seem any mirrored image should demonstrate some form of distortion.

If the concept of a luminous aether is overlapped onto current concepts of reflection of visible light, the explanation for reflection becomes seemingly more plausible. To conceptualize this, one must first adjust their perspective of the atom to that of an object comprised of a compact center and electron orbitals, buoyant in a three-dimensional universe comprised of quadsitrons. If white light is carried inbound by a transfer medium comprised of quadsitrons, and an atom, buoyant in the same three-dimensional medium, is struck by the inbound white light, the specific configuration of the atom generates a response in the form of vibration of the atom, which launches an outbound rippling effect into the surrounding three-dimensional medium, which represents a new wave of energy. See Figure 129. The frequency of the atom's vibration in response to the incoming energy wave is dependent upon the physical composition of the atom. The frequency of the response vibrates through the transfer medium to be witnessed by the human eye as the color of the object.

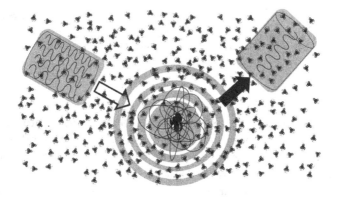

Figure 129

Inbound white light strikes an atom, causing the atom to vibrate in the three-dimensional luminous aether the atom is floating within, resulting in outbound light at a frequency specific to the characteristics of the atom.

The rippling effect of the three-dimensional medium or aether is in the form of visible light energy in one or more frequencies, which results in the object expressing a color or colors that the human eye can detect. See Figure 130. If the atoms comprising an object absorb all of the inbound energy without causing significant vibration of the atoms of the object, then the object will appear black. The absence of energy wave in of white light or an energy wave of a specific frequency in the visible light spectrum is interpreted as the color black by the human brain. Energy signatures of heat may radiate from an object which has absorbed a significant amount of energy, but the wave energy comprising the bandwidth of heat is not detectable by the human eye, but is detectable by the heat sensors positioned on human skin. If atoms comprising a mass vibrate at a specific frequency of the visible light spectrum, then an object appears to have color. If the atoms comprising a mass collectively vibrate with a majority of the frequencies in the visible light spectrum, the object will appear white to the human eye.

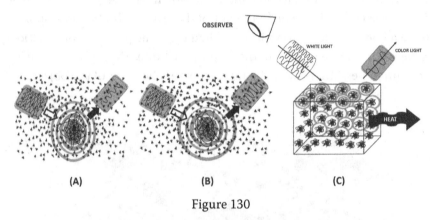

(A) (B) (C)

Figure 130

Inbound white light strikes a material, causing surface atoms to vibrate in the 3-dimensional medium, resulting in outbound light at a frequency specific to the characteristics of the oscillating atoms, to be seen by the eye. A & B: Differing arrangements of atoms or molecules, will cast the appearance of diverse color due to the vibration pattern being different. C: Color emitted by an object is related to surface atoms vibrating at one or more frequencies; heat is emitted by atoms/molecules emitting lower frequency energy.

To the astute reader, the existence of a rainbow seen as sunlight strikes a rain shower should cast doubt on a theory suggesting the presence of a luminous aether. As discussed above, the universe is three-dimensional, and therefore, as atoms vibrate, the effects of the vibration of atoms in relation to the surrounding luminous aether should be three-dimensional. Yet, as it has already been identified, a rainbow is typically seen by an observer when the observer is 42 degrees in relationship to the direction of the source of the light. Thus, it is a fact that visual detection of a rainbow is position dependent; would seem to contradict the concept of a luminous aether. It might be expected that if white light from the sun is vibrating atoms inside water molecules, then a rainbow should be appreciated from any angle of observation. The classical description of a rainbow, with no aether present, is light from the sun bounces off the water droplets comprising a rain shower. Again, at the subatomic level, specifically how is the behavior of light 'bouncing' off any object actually explained?

Paradoxically, instead of casting doubt, the phenomenon of a rainbow actually supports the existence of a luminous aether. The description of a rainbow moves the discussion of visible light to the next level, that of molecules vibrating to create specific color or range of colors. Molecules are comprised of two or more atoms. Various atoms combine per electron bonding to produce molecular structures.

A water molecule (H_2O) is comprised of one oxygen atom and two covalently bonded hydrogen atoms. See Figure 131. Oxygen has an atomic number 8 and standard atomic weight of 15.999. Each hydrogen atom has an atomic number of 1 and a standard atomic weight of 1.008. A standard droplet of water contains 1.67×10^{21} molecules of H_2O.[96]

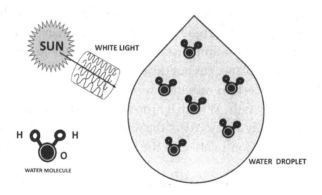

Figure 131

A water molecule is comprised of an oxygen
atom and two hydrogen atoms.

A water molecule is comprised of three atoms bonded together. Given the single oxygen atom and two hydrogen atoms are connected together, energy being absorbed by the individual atoms comprising a water molecule affects the behavior of the entire molecular structure. Conversely, energy released by the atoms comprising the water molecule is a result of the behavior of the entire water molecular structure. See Figure 132.

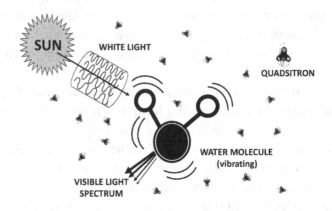

Figure 132

Given the atoms in a water molecule are bonded together, visible light energy emitted by the water molecule is preferentially directed in a specific direction by virtue of the construct of the water molecule.

Water molecules, by virtue of the two hydrogen atoms bonded to the single oxygen atom, exhibit a triangular like shape defined by the covalent bonds. In three-dimensional space, a water molecule may be symmetrical in one plane, but not the other two planes. More than likely gravity dictates a certain orientation of the water molecule when the molecular complex falls. Within a drop of water, during freefall such as in a rain storm, likely a majority of the water molecules align themselves in a similar orientation relative to the ground toward which the water droplet is falling.

Likely, the water molecule struck by white light absorbs the energy of the white light at the energy frequencies comprising the visible light spectrum. Then by virtue of the three-dimensional structure of the water molecule, the H_2O structure vibrates a majority of the energy that has been absorbed in a specific direction, specifically with regards to an observer positioned 42 degrees in relationship to the direction of sunlight. See Figure 133.

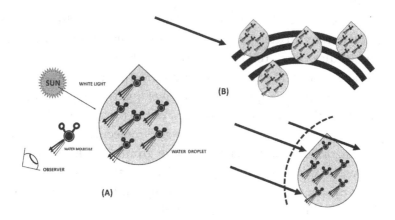

Figure 133

A. Struck by sunlight, the water molecules with in a water droplet all vibrate with in the luminous aether to create a spectrum of visible light directed in a specific direction.
B. Must consider entire rainbow as a three-dimensional object which absorbs sunlight.

It is likely, that if one were able to do an experiment onboard a spacecraft located outside the effects of any gravitational field, where white light was shown at a mist of water molecules in a confined space, the spectrum of visible light might be observed from all directions of observation rather than limited to 42 degrees in relationship to the direction of the source of the light.

When people dive in open water conditions, the further a scuba diver ventures below the surface of the water, the more color is lost. Red fades away first, then yellow, then green, then blue. At a depth of 130 feet the underwater surroundings appear a ghostly gray with almost no color. For dramatic underwater photography, a diver will generally bring his/her own light source to capture the vibrant colors sported by fish and other sea creatures at depth. So, a rainbow is similar. Red is on top, and blue is on the bottom. A cloud, which exhibits a rainbow, needs to be thought of as a composite of individual molecules of water, each water molecule falling through the air towards the ground, each water droplet acting as a collective prism, and then in addition, the cloud itself must be thought of as an entire entity, a three-dimensional body of water vapor. The water molecules comprising the cloud, like the ocean, absorb and re-vibrate the lower energy light waves first, while the higher energy penetrate further into the cloud to be absorbed by the deeper water molecules which re-vibrate the darker shades of blue light energy last. Exploring light mechanics beyond reflection and refraction leads to a rainbow acting as a splendid example of quadsitron-energy physics.

It is likely, the limitation to the view of the prism effect by sunlight striking a rain shower is due to the water molecules falling to earth in a similar orientation in relation to the ground. Without the effect of gravity, the water molecules in a mist of water would be oriented randomly in space and therefore resonating, emitting visible light and spectrums of light in various directions. In zero gravity, since the water molecules would not be oriented uniformly, the spectrum of colors cast by the water droplets may appear as be brief twinkles, rather than a fixed rainbow effect.

Last conceptual illustration regarding this subject: quanta of white light striking the molecules on the surface of an object producing color which can be detected by the human eye, is likened to a fixed iron bell being struck repeatedly by a linear stream of bullets. The result of a consistent stream of bullets hitting a fixed metal bell would be the production of a distinct sound and recurrence of the same audio signal vibrating through the air. Such a sound would continue until the bell was damaged, deformed, became dislodged from its position, or the firing of bullets at the bell ceased to occur. Quanta of light might be thought of as bullets striking an object. The result of quanta of light striking molecules is the generation of vibration of those molecules, like the bell would vibrate and create sound. The production of vibrations by molecules/atoms in the three-dimensional quadsitron field comprising the fabric of subatomic space then generates an energy signature, and if the vibrations are of energy with wavelengths in the visible light spectrum, then color is emitted into the subatomic quadsitron field and thus becomes detectable.

CHAPTER 33

HOW X-RAYS ARE GENERATED

WILHELM CONRAD ROENTGEN DISCOVERED x-rays in 1895.[97] While Roentgen, a German Physicist, was working with a cathode ray tube, he noticed fluorescent material in his lab was glowing a green color. His observations led to the discovery that an invisible form of energy was being emitted by his cathode ray tube, which was capable of penetrating human skin, but not bone or metal. Later in 1895, one of Roentgen's first x-rays, and most well-known radiograph, was that of his wife's left hand with a ring present on the fourth finger.[98] Quickly this observation was applied to study the human body and x-ray technology rapidly evolved given a newfound medical application.

The science behind generation of x-rays, limited by the constraints of the original technology, included evacuating a glass tube of most of its air to eliminate the nitrogen and oxygen molecules, and positioning a cathode electrode and an anode electrode within the glass tube. The original x-ray tubes were discharge tubes, with partial vacuum, invented by William Crookes (1832-1919) and referred to as Crooke's tubes.[99,100,101] When a high enough voltage was applied to the cathode, electrons would be emitted by the cathode; this process at times was referred to as boiling electrons off the cathode, though this is not necessarily an accurate description. It was thought that the emitted negatively charged electrons would cross the vacuum seeking the positively charged anode, a metal electrode acting as a receptor for the electrons. With the anode accepting electrons, current would flow if a closed electric circuit were constructed to include the anode and cathode. If the construct of the anode was of a proper metal, such as copper or tungsten, and if the voltage applied to the cathode were substantial enough, x-rays would be emitted from the anode as the electrons emitted by the cathode struck the anode, which was astutely observed by Wilhelm Roentgen in 1895.

The illustration of a crude cathode ray tube is provided in Figure 134(a). The figure shows a voltage applied to the cathode electrode, electrons crossing the space in the vacuum tube, striking the anode and generating a current in the output lead of the anode electrode. As electrons emitted by a high voltage catheter strikes a material such as copper, a carbonized filament or tungsten, the free electrons collide with electrons in orbit within atoms comprising a filament. The electrons striking each other cause the electron in orbit to be ejected. The free electron assumes an orbit at a higher level of energy around the nucleus of the atom. The collision of the two electrons causes energy to be released in the form of x-rays. See Figure 134 (b).

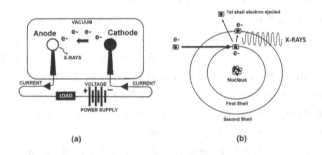

(a)   (b)

Figure 134

(a) Cathode ray tube used to generate x-rays, (b) x-rays generated by an electron in the first shell of a metal being struck by an outside electron.

The interesting observation is that if no luminous aether exists as the underlying fabric of the universe, then how is a sinus wave in the frequency spectrum of x-rays, generated by a free electron colliding with an electron in orbit around the nucleus of an atom; other than blindly accepting that such a phenomenon occurs without any necessary medium. When two pool balls strike on a billiard table, upon impact the first pool ball moves the second pool ball, but in addition sound waves are generated which our ears detect; such sound waves are sinus waves of audible frequencies, which are thought of as disturbing the air molecules which then transfer the sound energy to one's ears. Thus, if a medium comprised of quadsitrons indeed exists, then like billiard balls striking one another, the effect of one electron striking another electron would cause a vibration to occur in the medium surrounding the two electrons, which would create the sinus wave pattern to propagate through the luminous aether, which could be measured some distance away from the source as x-rays given a proper means of detection.

X-rays when directed at the human body, are only be partially absorbed, and only by the densest of tissues, such as bone, the remainder of the energy passing through the tissues. The radiation exiting the body, can be captured by specialized film, or most recently computer sensors, to create radiographic pictures of the biologic structures. The fact that x-rays pass through the human body to generate radiographs, is the preamble for the concept that the fabric of the universe, comprised of quadsitrons, also passes through the human body.

Sunlight, starlight, lightning, fire and chemical reactions generated by insects such as fireflies were the original, natural sources of visible light. Humans initially generated visible light by oxidative combustion of woods, oils, and petroleum products. In 1802, Humphry Davy (1778-1829) invented the first electric light using a platinum filament and an electric battery.[102,103] In 1850, Joseph Wilson Swan (1828-1914) created a successful 'light bulb' with a carbonized paper filament, which was successfully used to illuminate homes and public buildings.[104,105] Many versions of the light bulb followed. Thomas Edison, in 1880, was able to construct an incandescent light bulb which was commercially practical for widespread distribution to the public. Edison's bulbs had high vacuum, utilized a carbonized bamboo filament and had a high resistance which made them compatible with the construct of a public power grid.[106]

The incandescent bulb is generally constructed with a transparent glass housing. See Figure 135. Two electrodes are placed inside the glass housing. One electrode functions as a cathode, the source of electrons, while the second electrode acts as an anode, the receptor for electrons. Rather than the construct of a cathode ray tube, where electrons are required to cross an empty space between the cathode and anode, the incandescent bulb is constructed with a filament which bridges the gap between the cathode and anode.

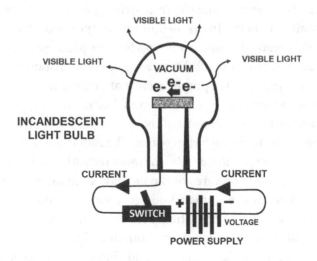

Figure 135

The incandescent light bulb emitting light in the visible light spectrum.

The filament connecting the two electrodes has been made of a variety of materials to include platinum, carbon arc, carbonized paper, powdered charcoal, carbonized bamboo, and most recently tungsten. Filaments were chosen due to their high melting point. A high voltage is applied to the cathode electrode, with this voltage being applied across the filament. The filament heats up. The vacuum prevents oxidation from occurring in the filament. The filament creates light. Less than 5% of the energy consumed by an incandescent bulb produces visible light, the remainder of the energy generated heat in near infrared wavelengths.[106,107] Incandescent bulbs fitted with a tungsten filament are capable of generating a continuous output of the visible white light spectrum.

Alternately, black light bulbs, are designed to emit ultraviolet-A light, and minimal visible light. Ultraviolet light is a longer wavelength than visible light, outside the visible light spectrum, and therefore not detectable by the human eye. Black light sources include fluorescent lamps, mercury vapor lamps, light emitting diodes (LEDs), lasers and specially constructed incandescent bulbs. Certain paints fluoresce under black light. The act of fluorescence makes the painted object visible to the human eye.

Still, in a universe where no luminous aether is recognized to exist, what a current flowing through a carbonized filament or a tungsten filament actually creates is difficult to conceptualize. The vacuum of an incandescent bulb demonstrates there is no interaction with the atmosphere in order to generate or transmit visible light energy, infrared heat energy or ultraviolet light energy when a voltage is applied to the electrodes of an incandescent light bulb. A student of the current day electromagnetic energy spectrum must just go on blind faith that sinus waves of energy of varying frequencies are created by a flow of electrons colliding with certain atoms and these same sinus waves are emitted into the environment beyond the boundaries of the atoms, without the benefit of some form of supportive transfer medium.

In a universe where a luminous aether exists, a student of the electromagnetic spectrum would conceptualize that a free electron would crash into an electron held in orbit about the nucleus of an atom in the filament, and the collision would generate a three-dimensional shock wave in the underlying surrounding fabric of quadsitrons. The vibration of one or both electrons involved in the collision would generate the sinus waves, which would then propagate out in all directions from the center of the site of collision, utilizing the quadsitron aether as the transfer medium. The frequencies of the sinus waves generated by the collision would be dependent upon the energy of the collision and the properties of the atom involved in the collision. The sinus waves, comprised of quanta of energy, would travel through the quadsitrons medium indefinitely until they encounter atoms which are capable of absorbing the energy.

In an incandescent light bulb, collisions of electrons or movement of electrons cause vibrations in the atoms comprising the filament, resulting in disturbance in the surrounding quadsitron field resulting in the emission of sine waves of the electromagnetic spectrum in the range of visible light, infrared and heat. The observation detected by humans is generally a form of white light and heat emitted by an incandescent light bulb. Light Emitting Diode (LED) light sources utilize a diode to generate light and emit much less heat than compared to an incandescent light bulb. A diode is a semiconductor device which typically facilitates the flow of current in one direction between to terminals. LED light may expend ten percent of energy required to light an equivalent incandescent light bulb.

CHAPTER 34

LUMINOUS AETHER FACILITATES LIGHT TRAVELING BILLIONS OF LIGHT YEARS

THE EXPLANATION OF HOW visible light travels from distant stars to the Earth to be viewed by those stargazers whom take the time to peer up at the night sky to study the celestial canvass, has remained a mystery. The most distant light emitting object thus far located is GN-z11, a high-redshift galaxy, 13.39 billion light year travel distance. GN-z11 is located in the Ursa Major constellation. [109]

Sir Isaac Newton wrote that light traveled long distances by means of a luminous aether. The presence of a luminous aether or some form of underlying fabric comprising the universe would make travel of light across the great expanses in the universe a plausible concept. At this time the explanation is that light just happens to travel millions of light years across empty space. If alternatively, a luminous aether existed, then if the quanta of light and the constituents of the luminous aether were both nonreducible, then, light would be forced to travel long distances across great expanses of space without dissipating, until absorbed by some molecular structure or diverted.

An energy source, such as a distant star, would create shock waves in the three-dimensional luminous aether, occupying the space surrounding the star, creating waves of white light and other frequencies of energy to radiate out from the star in all directions. See Figure 136. The origin of the shock waves is the result of vibration of the atoms comprising the star. The vibrations disperse, in multiple frequencies, as sine waves, in all directions away from the star.

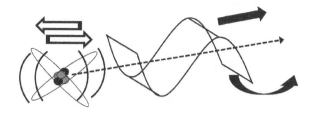

Figure 136

A distant star generates waves of energy in the visible light spectrum which ripple out from the source in all directions as a sine wave.

Given a luminous aether comprised of quadsitrons, atoms located at one location in the universe, vibrating the medium, may send out waves of quanta of energy toward distant locations in three dimensions. See Figure 137. Thus, the radiant energy in the visible light spectrum emitted by a distant star can be seen from all directions.

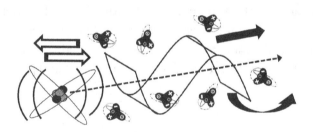

Figure 137

Quanta of energy radiate out from a light source
using quadsitrons as the transfer medium.

One of the critical concepts is that the luminous aether acts as a transfer medium in all directions in relation to the position of the original source of energy. See Figure 138.

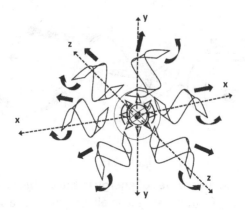

Figure 138

When a star emits light, the energy ripples out
from the source in three dimensions.

Light ripples out from the light source of a distant sun. White light emitted by a star is comprised of numerous frequencies of energy waves within the narrow band of visible light. See Figure 139. Visible light is the bandwidth of frequencies of 430-770 THz (terahertz (10^{12})). Stars will generally emit energy waves of other frequencies within the electromagnetic spectrum, but the visible light spectrum is the bandwidth of energy the human eye is able to detect and process as part of human vision.

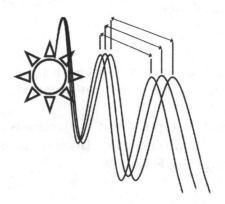

Figure 139

A star emits numerous energy waves of various
frequencies of the electromagnetic energy.

Since quanta of light are nonreducible and the quadsitrons comprising the transfer medium in space are nonreducible, the energy emitted by a distant sun is forced to continue in its travel through space undisturbed until the quanta of energy encounter a substance which will absorb, trap or divert the course of travel of the quanta. See Figure 140.

Figure 140

Given quanta of energy and quadsitrons are nonreducible waves of energy transfer through space intact until absorbed or diverted.

Quanta of energy comprising a wave of energy eventually encounter an object comprised of atoms, which will absorb the quanta of energy. Such an object may be the human eye. See Figure 141. A human, peering up into the night sky, will detect stars dotting the black canvass. The white light generated by the distant stars has traveled many light years to arrive at earth, where the retina of the human eye is capable of absorbing the white light and the human brain is able to interpret the visible light from the distant suns as a stars.

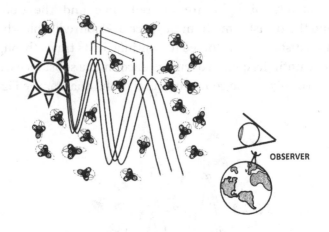

Figure 141

The human eye absorbs an energy wave that has
traveled across space generated by a distant star.

The atoms of a star vibrate causing waves of energy to be emitted by a star. The waves of energy are a sphere of energy as light waves generated by a star leave the star, traveling in all directions in three-dimensional space. As light travels away from a star the radius of the sphere progressively increases per the standard equation for surface area of a sphere which is Spherical Surface Area = $4\pi R^2$. As the radius of the sphere increases the surface area of the sphere increases per the equation provided. As light travels away from the star of origin, the three-dimensional surface area light must occupy becomes progressively larger. In response, the amplitude of the original three-dimensional light wave diminishes. The frequency of the signal does not change and the individual quanta of light comprising the amplitude do not diminish. The number of quanta headed in a particular direction reduces as the light is spread over a continuously increasing three-dimensional surface area of space. See Figure 142.

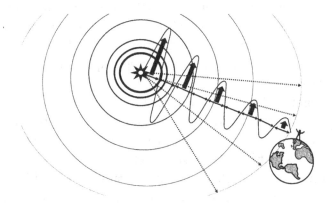

Figure 142

The amplitude of the wave migrating through space diminishes
as the distance from the original light source increases.

The existence of a transfer medium such as quadsitrons facilitates
that though the quanta of light reduce in amplitude as a light wave
distances itself from its source, visible light of all frequencies travel from
one location in the universe to another location simultaneously, to arrive
at the same location at the same time. See Figure 143.

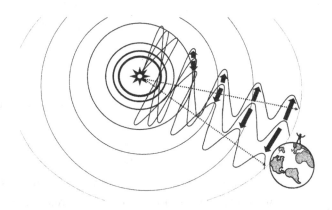

Figure 143

The quadsitron transfer medium, this luminous aether, facilitates
visible light of all frequencies to travel across the universe
to arrive at the eyes of a stargazer at the same time.

The stargazer standing on the Earth, observes white light being emitted from a distance sun which is capable of radiating light from the visible light spectrum. Some stars may only be able to emit light of limited bandwidth creating stars which might appear red or blue to the human observer.

White light emitted from a distant star, traveling as white light, transits the vast distance to the earth and generally appears as white light to an observer standing on the Earth. See Figure 144. That is, when one looks up at a night sky, the stars tend to twinkle due to fluctuations in earth's atmosphere, but the star light generally appears white; the light from distant stars does not demonstrate a spectrum of visible light, that is changing from red to blue to violet to yellow to other colors in the visible light spectrum.

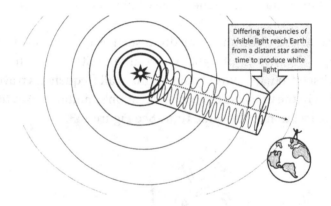

Figure 144

White light arrives to the earth from a distant sun radiating white light.

If there were no luminous aether between distant stars and the earth, then, since light travels in a wave, light from different frequencies in the visible light spectrum would have to travel different distances depending upon the frequency of color the light. Since the speed of light, as light traverses through space is considered a constant by most scientists, if the length of the path taken by light changes then the time of travel from one point in space to another point in space would change. Given the differing frequencies of the visible light spectrum, one would expect that different colors of light would arrive from a distant star at differing times to an observer standing on the earth. If there is no aether in space, an observer peering up at the night sky should view all of the stars radiating white light to change color over the spectrum of visible light. See Figure 145.

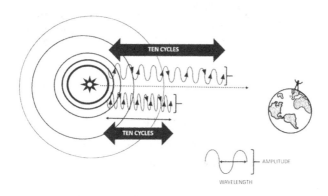

Figure 145

Variance is distance traveled across space related to variance in frequency of visible light spectrum. Different colors of light would arrive at the earth at differing times due to the imbalance in distance each frequency of light would need to travel from different star to the earth.

One might argue that white light would be the end result that an observer would see due to the magnitude of light emitted by a distant sun. This argument might be true of stars closest to the earth, such as our own sun. But stars may be hundreds or even billions of light years away, and with such a vast distance of travel and such low magnitude, that it would be expected that such distant stars would change color due to the discrepancy of the distance the various frequencies of light traveled to reach the earth.

If there exists a luminous aether comprising all space in the universe, the scenario changes. The aether creates a means for white light, emitted by a distant star, to traverse from one point in space to another point in space and arrive at the destination with all frequencies of visible light arriving at the same time. See Figure 146.

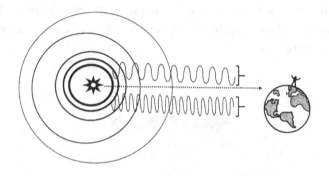

Figure 146

All frequencies of white light arrive from a distant
star to the earth simultaneously.

The existence of a luminous aether makes it possible for all
frequencies of visible light to reach an observer on earth at the same
time. The presence of an aether, comprised of quadsitrons, creates a
means for the energy transfer of white light emitted from a distant star
to utilize the medium to transit from one point in space to a second point
in space and have all frequencies arrive at the same point, such as the
earth, simultaneously. See Figure 147.

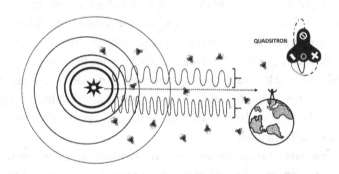

Figure 147

All frequencies of white light arrive from a distant star to the earth
simultaneously due to the presence of an aether comprised of quadsitrons.

The concept of light traveling through an aether is similar to the
concept of sound waves traveling through water. Generally sound waves
travel at a speed of 1,482 meters per second through fresh water, and

343 meters per second through air, both at 20 degrees celceius.[110,111] The speed of sound traveling through water is dependent upon density and temperature of the water. If a noise, comprised of differing frequencies, is created, then transmitted through a body of water, all of the frequencies of the sound utilize the water as their transfer medium, to arrive at the point of sound detection at the same time. Distortion to the sound may occur due to bubbles, impurities in the water, changing density or changing temperature of the water. The various frequencies of a sound utilize the medium the same to transit from location of origin to the location of detection under water.

If light did not travel through a luminous aether from point of origin of white light at a distant star to point of observer located on earth, then the stars in the night sky should be twinkling differing colors. Given the spectrum of white light includes deep blue and purple, the stars should be appearing to fade in an out of perception as the deep blue and purple hues against the black background of the night sky, would be observed by a stargazer located on the earth. Stars twinkle, due to changes in the atmosphere surrounding the earth. A star might emit light of a particular color frequency, such as a red dwarf star emitting red light. Generally, the stars which light up the night sky appear to emit white light, suggesting all of the frequencies of visible light emitted by a star reach the earth simultaneously.

Similar to sound traveling through water, light traveling through a luminous aether comprised of quadsitrons, may be dependent upon and thus become distorted from its original form, due to density of the quadsitrons, bubbles in the aether and directional flow of the quadsitrons.

CHAPTER 35

RE-FOCUSING THE OPTICAL THEORY AS RELATED TO A LUMINOUS AETHER

VISIBLE LIGHT IS A subset of wave frequencies of energy in the electromagnetic spectrum in the frequency range of 430-770 THz. Visible light, like the other electromagnetic spectrum moves through space in a wave pattern comprised of individual photons of energy. See Figure 148.

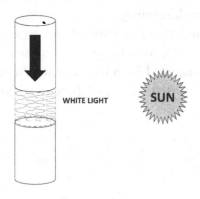

Figure 148

Visible light transits through space as waves. White light is comprised of the entire spectrum of visible light.

Albert Einstein described a universe constructed of mass, gravity energy. He taught that massive bodies in the universe, such as stars and large planets, would act to warp gravity and time.[112,113,114] Einstein predicted that the warping of gravity by celestial bodies of large mass would change the trajectory of light, as light transited through space. He predicted that light, passing by a heavenly body exhibiting an intense gravitational field would be drawn toward such a heavenly body due to the warping of the gravity field in the space surrounding the celestial body. See Figure 149. Following Einstein's prediction, this phenomenon was proven. Photos were taken before and during a total solar eclipse by British astronomer Sir Arthur Eddington in 1919, which demonstrated that light from distant stars, did indeed migrate closer to the sun as visible light passed by the sun at the center of our solar system.[115]

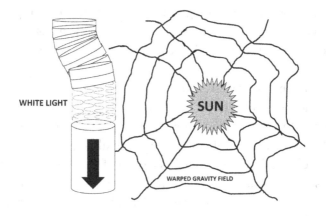

Figure 149

Theory: Visible light passing by a gravitational body has been shown to be drawn toward such a body is hypothesized to be due to the warping of the gravity field.

Thus, the phenomenon of light being drawn toward a gravitational body is not a contested concept. Proof of Einstein's prediction did result in an upheaval in the previous well-established century old Newtonian physics, and made way for quantum physics to take center stage as the mainstream dogma of physics.[116,117] Quantum physics was based on the concept that energy traversed from one location in the universe to another location, in the form of discrete packets, referred to as 'quanta'.

Max Planck in 1900, recognized energy traveled in multiples of a specific unit. He determined the unit was 6.62607004 x 10^{-32} m^2 kg/s, which has been referred to as Planck's constant. Max Planck developed the equation $E = h * v$, where E is energy, h is Planck's constant and v represents frequency of radiation.

To compile his signature equation, $E = MC^2$, Albert Einstein designated 'C', representing the speed of light as a constant. Thus, the visible light passing by a heavenly gravitational body, should pass by such a body at the designated speed of 720 million miles an hour. Thus, a paradox is established: if light has a constant speed, then neither gravity nor any other force should be capable of affecting the course of speed of light. The paradox is further enhanced by the theory that a black hole traps light. Again, if light has a constant speed, how can any force, regardless of how strong the force might be, halt the forward progress of light, trapping the electromagnetic energy. Light, traveling at the speed of light, is not expected to possess any appreciable mass, thus the means by which gravity is able to alter the course of a massless entity remains to be explained.

Albert Einstein received the Nobel prize in physics in 1921 for his Law of Photoelectric effect. Einstein was not able to attend the Nobel prize presentation ceremony. Instead, he provided a lecture to the Nordic Assembly of Naturalists at Gothenburg, on July 11, 1923. Since the lecture was not delivered on the official occasion of the Nobel Prize award, the contents of the lecture was not compulsory regarding the topic of the law of photoelectric effect for which the Nobel Prize was awarded. Einstein instead, directed his remarks towards further elaborating on his Theory of Relativity.[118] During the course of the delivery of the lecture Einstein spoke negatively towards the existence of a luminous aether, since the existence of an underlying fabric to the universe would have clashed with the Special Theory of Relativity.

Contributing to the paradox, there has not been a substantial explanation for the existence of gravity. Gravity is thought to be the term for a force which attracts or pulls objects towards the center of a planet or a star or a black hole. In addition, there has not been a viable explanation as to how gravity would affect light. It has been theorized, that light is representative of pure energy.

A paramount concept which (1) could explain the paradox of light appearing to be drawn closer to a heavenly body exerting a strong gravitational force, and (2) would not sacrifice the concept that the speed of light is a constant, would be the existence of a luminous aether. In a universe predicated by a luminous aether, quadsitrons would be packed closer together when near a heavenly body exerting a gravitational force versus farther away from a heavenly body. See Figure 150. The pumping action of the core of a sun would naturally cause quadsitrons to be more compact closer to a sun, rather than more distant to the sun. It is reasonable to concur that if the flow of wave energy is dependent upon the existence of quadsitrons as the transfer medium, and therefore it is also reasonable to assume that energy would follow the path of least resistance. If the quadsitrons are denser nearer to a gravitational body, then light would maintain its speed, but divert its course closer to the sun as Einstein predicted, to take advantage of the easier more uniform and consistent means of travel. A similar effect is observed in an electric circuit where current is flowing, if there are alternate paths with which the current may flow, the electrons will tend to take the path of least resistance.

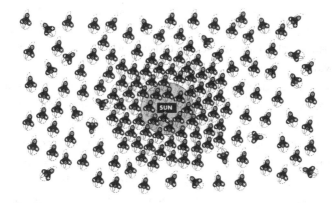

Figure 150

Quadsitrons are drawn closer to celestial bodies, which have a molten core or other form of quadsitrons pump; thus, the number of quadsitrons becomes denser the closer the position is to a star.

Humans would tend to behave in a similar manner. If one were to travel by car from destination A to destination B, and were required to maintain as close as possible a constant speed, and if given a choice between a bumpy gravel road and a smooth paved concrete road, most drivers would choose the paved road; unless possibly if they were on holiday and had a desire to see a more rural landscape. See Figure 151.

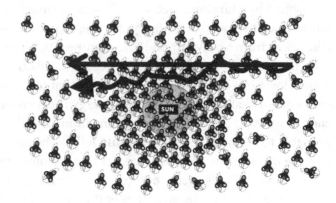

Figure 151

The density of quadsitrons is higher near a sun, therefore light would
travel closer to the sun to take advantage of the higher density
of quadsitrons resulting in a more uniform transfer medium.

Alternate explanatory concept: if crossing a stream of water by
hopping stones, one would generally choose to hop across stones closer
together, rather than stones separated farther apart; as to avoid falling
into the water.

We have arrived at the moment of truth. Factually, do quadsitrons
exist? More explicitly, does a luminous aether comprised of quadsitrons
realistically act as the underlying fabric of the universe? A simple
experiment could be contrived that could support the existence of
quadsitrons representing the luminous aether. The experiment would
be designed as a modified version of Einstein's prediction, which was
that light would travel closer to a heavenly body exhibiting a strong
gravitational field.

If there is no luminous aether, since white light is comprised of
varying frequencies, if the light were to be attracted toward a gravitational
body by an unexplained force of 'gravity', then it is conceivable that the
width of the white light beam would widen due to the gravitational
pull exhibited by the sun, producing an effect similar to what is seen
when white light passes through a prism. That is, the attractive effect
of gravity would physically spread out a beam of white light, putting on
display the varying colors of the visible light spectrum as the beam of
light physically widened. The pull of gravity from the sun would have a
differing effect on the various energy levels of the differing frequencies

of the colors of light comprising white light, which would result in the spread of the beam of white light. See Figure 152. A beam of white light transmitted by a distant star would appear as a spectrum of colors, as the beam of white light passed by the sun.

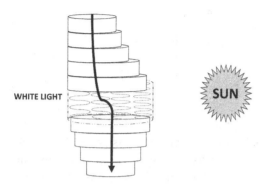

WHITE LIGHT

SUN

Figure 152

If no luminous aether exits, white light would be transmitted
by a distant star would appear as a spectrum of colors,
as the beam of white light passed by the sun.

If, alternatively, a luminous aether does exist, as Sir Isaac Newton believed, then a beam of white light would remain intact, appearing as white light, as it passed by the sun. See Figure 153. The varying frequencies comprising white light would utilize the density of the quadsitrons equally, as the beam of white light passed by the sun, thus there would be no spreading of the white light into a spectrum of varying colors.

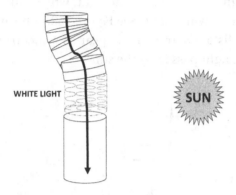

Figure 153

If a luminous aether does exist, as Sir Isaac Newton
believed, then a beam of white light would remain intact,
appearing as white light, as it passed by the sun.

The observational experiment which proved light bends towards the sun was done during a solar eclipse. If the same experiment is performed, but utilizing today's improved optics technology to detect a spectral change in white light as the light of a distant star passes by the sun, if a luminous aether exists, then the beam of white light will remain intact as white light; if a luminous aether does not exist, then the beam of white light from a distant star will show a spectrum of color as it is shown to move closer to our sun on its way to the earth.

Unfortunately, though such an experiment may, at first, sound plausible, the pictures would have to be taken near the sun, rather than from a location on the earth. While light from a distant star passes by the sun, gravity would have its most intense effect on a beam of electromagnetic radiation, when the light is closes to the sun. As the beam of light originating from a distant star continues past the sun, the more distance between the sun and the light, the less effect the sun's gravity would have on the beam of light. At a sufficient distance from the sun, as the beam of light is headed toward the earth, the sun's gravity would diminish to the point of having little effect on the beam of light, and thus the widened beam of light would possibly contract back to the original width. If the widened beam contracts to its original width as the light reaches observers positioned on the earth, then the spectrum of colors which might be seen near the sun, would collapse back to a beam of white light; which would be recognized on earth as the normal image of a star.

The astute student of photoelectric effect then should ponder two questions regarding the above claim. The problem with the above description is that if a beam of light passing by the sun, were to be physically spread out such that the white light now took on the image of a spectrum of colors, by what mechanism would the now widened beam collapse back into a thinner beam. It is doubtful, that light, an entity with no mass, has a spring like mechanism, which would cause a distorted beam of light to resume its original shape without the benefit of an outside force. Thus, this challenging observation would suggest that if Einstein's hypothesis regarding gravity and light were true, light from a distant star passing in close proximity to the sun would be observed as a spectrum of colors, rather than a beam of white light.

The second question the astute reader would ask themselves, is that Albert Einstein envisioned, the photographer Sir Arthur Eddington who took the photos of the solar eclipse in 1911 envisioned and most students envision that light from a distant star is received on earth in the form of a pencil beam. We see an illuminated dot in the night sky, which we refer to as a star and we observe from a single observational point on earth such light to be a star. The human brain interprets the light from a distant star as a pencil beam. Since from any observational point on earth where the light of the star might be witnessed, the star is observed, the light from the distant star is obviously not constructed as a pencil beam, like one might see with a laser, otherwise other observation posts where the same star can be viewed would not be able to see the star. Since the same star can be viewed from various points in three-dimensional space, the light emitted by a distant star is in the form of a sphere, not in the form of a pencil beam. This concept of light originating at a specific point, such as a star and propagating through space as a wave with the geometry of a sphere, was previously discussed in Chapter 34 and illustrated in Figures 136-147.

Thus, if light originating from a radiant energy source, travels through space as a wave with the geometry of a sphere, light is a three-dimensional entity, rather than a two-dimensional pencil beam entity. A three-dimensional wave energy passing by the sun still would comply with the above concept that light would preferentially follow a path of denser quadsitrons, thus deviate toward a gravitational body such as the sun on a flight path past the sun. The fact that light is a three-dimensional wave energy probably suggests that the above claim that light would possibly separate into a spectrum of colors if no luminous aether exists, is probably an invalid claim. Whether there is a luminous aether or there is not a luminous aether comprising the universe, light still is a three-dimensional wave energy entity and would likely appear the same in both instances under the same sampling condition. The point of this mental exercise was to disavow the pencil beam light wave concept and open the analytical mind to the three-dimensional light wave energy concept.

QUADSITRON MECHANICS: HOW THE UNIVERSE IS CONSTRUCTED

CHAPTER 36

BLACK HOLE: VORTEX OF A GALACTIC HURRICANE

CURRENTLY ACCEPTED DESCRIPTION OF a black hole is this phenomenon represents a region in space where gravity acts on light so intensely, light is not able to escape such a region in space, and thus the name black hole.[119] What plagues the validity of the currently accepted description of a black hole is that neither gravity is explained, nor is the means by which gravity exerts a force on light explained. An attempted explanation is that the origin of a black hole is the implosion of a star, such that the matter comprising the original star collapses in on itself, creating an intense gravitational field, the force generated by the collapse of matter so powerful nothing is able to escape such a region in space.

The above explanation of a star collapsing in on itself creating an intense gravitational field creates several paradoxical concepts which are difficult to reconcile. Collapse of a star involving reduction of atomic structures imploding to a substance comprised of simply protons, neutrons and electrons without space between the nucleus of the atom and the electrons would seem to result in a dramatic loss of energy rather than an increase in energy. The primary means atoms harbor energy with in their structure is through function of the electron cloud which surrounds the nucleus of the atom. To eliminate the electron cloud, would seem to result in loss of energy storage capacity of an atom, not an increase in energy as surmised by the collapse of a star. A similar analogy is the wood of a tree being burned in a fire. Wooden logs are much larger in volume than the resultant ash created when a log has been burned. Wood, in the state of ash as a result of burning, has exhausted much of its energy during the process of combustion. Far less energy is present in the residual ash left at the end of a fire, than the wood logs which are burned to create the fire. Thus, if the atoms of a star were spent in the process of combustion and the atoms collapsed to the point the electrons had lost the structure of their orbitals, then the resultant condensed star would seemingly possess very little residual energy, contrary to the postulated increase in energy.

The presence of a luminous aether explains the phenomena of a black hole more pointedly than the current, simplified explanation that a black hole is the result of the collapse of a star's atomic structures. The presence of a luminous aether filling three-dimensional space spanning the entire extent of the universe provides an elegant explanation of black holes and other celestial phenomena. A vantage point for the study of black holes present on earth would be that of being underwater as a scuba diver surrounded by water, without the influence of gravity from earth's core. The perspective one requires for the study of celestial phenomena, such as black holes, is in a three-dimensional medium surrounding and suspending the student, on all sides, in all directions.

The earth's sun is located in the Orion spur of the Milky Way Galaxy.[120] See Figure 154. The Orion spur is a grouping of stars that act as a curved arm in the spiral galaxy. Like most spiral galaxies, the black hole is located at the center of the galaxy, with the spirals swirling and radiating outward from the center.

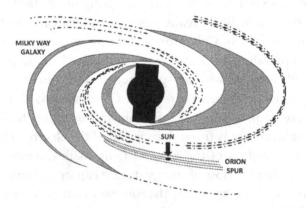

Figure 154

Milky Way Galaxy (top view), with the Sun located in the Orion Spur.

In a universe where a luminous aether exists, and therefore an essential nonreducible entity exists to create the aether, one would expect such an aether to be mobile.

The Milky Way galaxy contains a supermassive black hole at Sagittarius A, a very bright star and astronomical radio source at the center of the galaxy. Viewed from Earth, Sagittarius A is located between the constellations Sagittarius and Scorpius. Sagittarius A was discovered in 1974 with an interferometer at the National Radio Astronomy Observatory.[121]

Similar to water, the luminous aether would behave with currents, flows, swirls, funnels, and waves as the aether continuously pursues to evenly distribute the essential elements to enable balance in the flow of energy that traverses the universe. Given the energy flowing through the universe can never be in absolute balance, the movement of quadsitrons as a mass fabric of the universe is always dynamic and in flux.

A spiral galaxy, such as the Milky Way galaxy, is thus a hurricane-like cloud comprised of a mix of both free quadsitrons and quadsitrons bound by energy to create neutrinos, protons, neutrons, dark mater, atoms, and molecules, continuously swirling around a central point. See Figure 155. At the center of a spiral galaxy is a vortex. Surrounding the central vortex are clouds of stars arranged in wispy arms that churn around the center axis of the galaxy.

On the earth, dense collections of air accompanied by water vapor and debris are regularly observed around the world as tornados, water spouts, hurricanes and typhons.

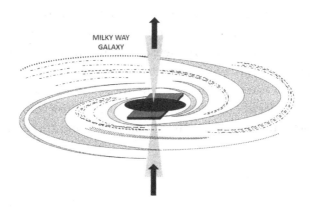

Figure 155

Spiral galaxy seen from the perspective of the side view.

The central vortex of a spiral galaxy twists and compresses the material of a galaxy, much like the funnel of a tornado seen in earth's atmosphere. See Figure 156. The illustration seen in Figure 156 is that of the vortex seen on its side. There exists an event horizon on both sides of the vortex. An event horizon refers to the opening on either end of the vortex where materials and energy enter one end of the vortex and exit from the opposing end of the vortex.

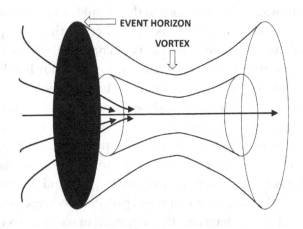

Figure 156

The vortex of a spiral galaxy with an event horizon present on both sides of the vortex.

Quadsitrons line up on the side walls of the vortex. See Figure 157. The walls define the side boundaries of the central vortex, similar to a twisting tornado having walls of dense air defining the boundaries of the wind tunnel of the tornado. Similar to a tornado wrenching and pulling objects up from the ground where the tornado's vortex touches down on land, the center of a spiral galaxy consumes matter into its vortex on one side and emits matter out the opposite side into three-dimensional space.

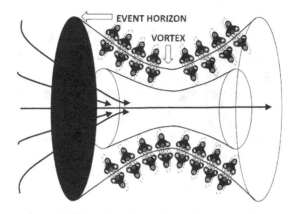

Figure 157

Quadsitrons line up on the sides of the spiral galaxy's vortex.

Spiraling occurs inside the vortex. See Figure 158.

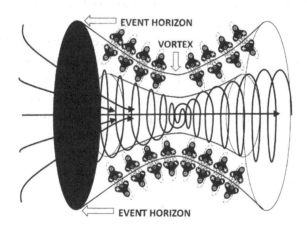

Figure 158

Spiraling occurs inside the vortex of a spiral galaxy.

Given the spiraling which occurs inside the vortex of a spiral galaxy, a differential occurs between the two event horizons. One event horizon will suck quadsitrons into the interior of the vortex, while the high intensity spiraling at the center of the vortex will churn the quadsitrons and force the quadsitrons out the opposite end of the vortex. See Figure 159.

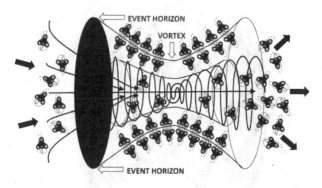

Figure 159

Quadsitrons are sucked into the vortex at one event horizon and ejected out the opposite event horizon of the same vortex.

Light follows the behavior of its transport medium as light courses through the universe. Quadsitrons act as the transfer medium for light. As quadsitrons flow into the event horizon and become influenced by the twisting funnel of the vortex of a galaxy, light following the flow of quadsitrons also enters the galactic vortex. See Figure 160. The action of light flowing into a vortex creates the appearance of a black hole when viewed from the side of the spiral galaxy from which light enters the vortex; no light is reflected from this point and no light can be transmitted through this region of space from the opposing side of the galaxy, thus this region of space appears void of light; that is it would appear as a black hole.

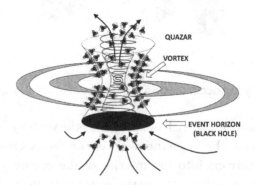

Figure 160

Light follows its transfer medium into the vortex of a galaxy, creating the appearance of a black hole.

Light passes through the vortex following the flow of quadsitrons. The quadsitrons are ejected out of the opposing side of the spiral galaxy by the churning force of the vortex. Light continues to follow the behavior of its transfer medium and likewise bursts out the opposite end of the spiral galaxy's vortex along with the quadsitrons, which are being ejected. Viewing the spiral galaxy from the side where the quadsitrons are being ejected gives the appearance of a region of intense brilliant white light located at the center of the spiral galaxy. Such spots of intensely bright light are often referred to as a quasar.

Quasars are thought to represent some of the brightest, most intense regions of space.[122] Given pure electromagnetic energy spews forth from a vortex at the center of a spiral galaxy, would create locations dotting the universe, where when viewed from the side of a spiral galaxy where the light is being emitted, would be intensely bright to a human observer. The vortex emits many frequencies of electromagnetic energy and in addition to white light, a vortex would spew out gamma rays, x-rays, radio waves, and microwaves. The various twisting motions of differing galaxies may lead to preferential frequencies being generated by the vortex of various galaxies.

Viewing a spiral galaxy from the side with the axis of the vortex lined up from top to bottom, and incorporating the behavior of the vortex of the galaxy, creates a similar picture as a bar magnet. See Figure 161. A galaxy may have a north pole and south pole. In part, the energy flowing through the vortex of the spiral galaxy and emerging from the quasar is channeled up from the vortex, then flows around the outer boundaries of the galaxy to again, in part, return being sucked into the black hole of the galaxy. The spiraling nature of the spiral galaxy creates a massive three-dimensional spherical force as the galaxy churns through space.

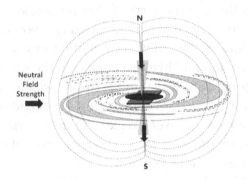

Figure 161

Spiral galaxy demonstrates a north and south pole.

The vortex of a spiral galaxy commands the most intense forces possible in the universe. The twisting and churning motion which occurs at the mid portion of the vortex is so intense, it is capable of disassembling matter. Matter is comprised of atoms. Atoms are comprised of protons, neutrons and electrons. It is possible at the mid portion of the vortex of a very powerful spiral galaxy, not only are molecules ripped apart, but atomic bonds internal to atoms may be broken and the construct of atoms may be torn apart down to the very elements of protons and electrons, possibly broken down to the very primary elements of matter the quadsitrons. See Figure 162. Given the intense pressure at the center of the vortex, the swirling mixture of quadsitrons and pure quanta of energy, some fused into neutrinos, emerges from the caldron of the vortex, recreating protons and neutrons, then recreating atoms, and finally atoms are merged into molecules.

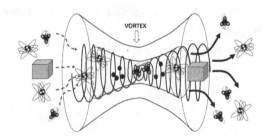

Figure 162

The vortex of a spiral galaxy disassembles matter, actively renewing the elements of the universe.

The vortex of a galaxy may act to continuously renew the elements of the universe. The intense forces generated by the vortex of some spiral galaxies make such a galaxy capable of reducing matter to the primary element of the quadsitrons. Quanta of energy and quadsitrons cannot be reduced, only rearranged. Renewed molecular mass is ejected by the vortex back out into space. The vortex of galaxies dynamically and perpetually reorganizes the matter and energy comprising the universe.

CHAPTER 37

SECOND TYPE OF BLACK HOLE

A BLACK HOLE IS a phenomenon recognized as a region of space where light does not escape, thus the name 'black' hole. It has been thought that a section of space where a black hole exists, represents a region where the gravitational forces were so excessive that gravity affected the travel of light to the point a black hole actively traps light. It has been long thought that light was not capable of escaping the gripping forces of gravity. This contradicted the concept of light as used in the equation E = MC², where light is said to represent a constant speed of 299, 792, 458 m/s or 720 million miles an hour.[38]

What has generally been thought is that light represents energy and does not represent matter. If light does not represent matter, then the question has always been if gravity does interact and affect light, how would gravity affect an entity that does not consist of matter.

The previous chapter discussed the concept that light would enter into a vortex of spiraling quadsitrons and light would follow the flow of quadsitrons through the vortex to the opposite side of the vortex. The center of a spiral galaxy is comprised of a vortex. Therefore, view a spiral galaxy from a distance would demonstrate a black hole in the center of the spiral galaxy where the light entered the vortex; the opposite end of the vortex would represent a quasar where the light would be escaping into the surrounding space. The vortex in the center of a spiral galaxy would appear black, not because light would be trapped in the black hole, but instead due to light following the path of its transfer medium as light disappeared into the vortex on its way to emerge on the other side of the spiral galaxy.

This description opens up the possibility of a second type of black hole in existence throughout the galaxy. Light is dependent upon the existence of a density of quadsitrons to fill a space between point A and point B in space. Actually, since space is three-dimensional, light is dependent upon the existence of, in effect, a cloud of quadsitrons filling the space between two locations in space represented by the universal three-dimensional coordinates of point A (x_1,y_1,z_1) and of point B (x_2,y_2,z_2), with x, y and z representing the three axes in space.

The second type of black hole exists where there is an absence of quadsitrons. If there exists an insufficient amount of quadsitrons between point A (x_1,y_1,z_1) and of point B (x_2,y_2,z_2) in space, then light is unable to pass from point A to point B through such a region due to a lack of sufficient transfer medium to make the passage between the two locations possible. See Figure 163. Such a region of space might be likened to a bubble in space. A bubble suspended in a body of water is a location where generally there is a pocket of air, but no water molecules. Similarly, the universe may be dotted with locations where quadsitrons do not exist, in effect creating a bubble.

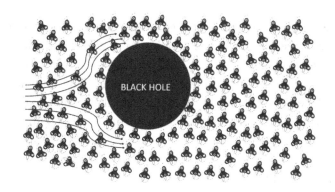

Figure 163

A second form of black hole can be created by a region in space which lacks the presence of quadsitrons, thus light is incapable of passage through such a region.

Thus, in locations around the universe were gravity exists at such an intense density as to displace quadsitrons from a region of space, this region of space will appear black or absent of light due to light not being able to pass through such a region. In the case of this second type of black hole, the light will divert around the black hole, making the perimeter of the black hole illuminate due to the light that would have traversed the black hole, having traveled around the boundaries of the black hole. See Figure 164. From an observer's stand point, one would see the light around the edges of the black hole.

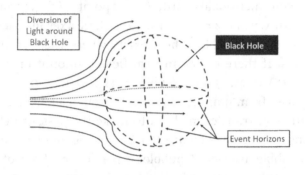

Figure 164

Light diverges around the perimeter of a region in
space where quadsitrons do not exist.

With gravity affecting the flow of quadsitrons, rather than the flow of light, the concept that the speed of light remains a constant remains intact. The question as to whether light could move faster than the speed of light if the transfer medium itself was capable of moving faster than light and if when quadsitrons move faster than the speed of light, can the quadsitrons pull light with themselves or is the speed of light a constant no matter how fast or slow the quadsitrons are moving. Most likely, the speed of light is a constant, independent and irrespective of the speed of the transfer medium.

CHAPTER 38

DENSELY PACKED QUADSITRONS, 3RD TYPE OF BLACK HOLE

IN THE MICHELSON AND Morley era, which is current state of physics, there has existed a fundamental unexplained, unreconcilable observation in physics; that of the two differing and conflicting behaviors of light. In some circumstances light behaves as a wave, and in other circumstances light seems to exhibit the quality of a solid. The dilemma is, that current theory considers that a substance moving at the speed of light is not supposed to exhibit the property of mass, only the characteristics of pure energy.

In the Michelson and Morley era of physics, with the notion of the existence of a luminous ether discredited, it is difficult to explain the two behaviors of light. If a luminous aether does exist, then the behavior of light as a wave and the behavior of light as something with mass is explainable. The aether provides the means of explanation.

Water easily and transparently exists in the state of vapor, liquid and solid before our eyes in the environment we occupy. Given the density of water, which is driven by the amount of energy which excites the water molecules, water can act as a wave or if cold enough, the same water can solidify and act as a solid. The most pointed example of a solid would be a snowball one forms in their hands and playfully throws in the direction of a friend on a cold winter's day.

Reviewing quadsitrons as the transfer medium for light, given the density of the quadsitrons in a particular volume of space, light would behave as a wave, or with regards to increased density of quadsitrons, light may behave more as a solid.

Similarly, sound waves easily transfer through air. Sound will transfer through water, but less coherently and for a shorter distance. Sounds transfers through the denser form of water, when frozen, but less coherently. Packing water molecules tighter together into a more rigid substance, such as ice, impedes the transfer of sound energy through a medium comprised of water molecules.

Volumetric regions of space where an energy field is so intense that it compacts the quadsitrons into a highly dense configuration, leads to the third type of black hole, that of a densely packed quadsitron field. Similar to the observation that sound does not permeate and travel through ice as easily as air or liquid water, likewise, light does not permeate or travel through regions of quadsitrons which are so densely packed together by an external compressive force field that locks the quadsitrons in place in relationship to each other such that they are not independently capable of movement. See Figure 165. Quadsitrons, held rigid, and not capable of any measure of movement, are made incapable of acting as a medium for light to transfer from one location in space to another location in space.

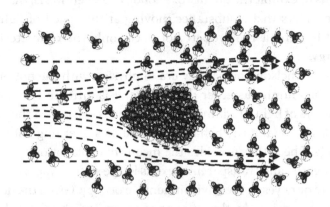

Figure 165

Intensely packed quadsitrons are incapable of acting as a transfer medium for light, thus creating a region in space that will act as black hole.

Due to the four charge states, the spherical 120 degrees between each of the four charged poles, and the energy fields each quadsitron emits, quadsitrons generally do not become locked in place. Quadsitrons typically maintain a minute space between each sub-sub atomic entity and slide across each other. It would be exceedingly rare for quadsitrons to lock up into a sub atomic solid structure where the internal sub-sub atomic entities so not move independent of each other. An immensely strong gravitational force would be required to pack quadsitrons together such that they could not rotate in space on their own.

Light, as all forms of the energy in the electromagnetic spectrum, must continue to move or flow in some fashion. Preferentially, light will follow a path of least resistance and seek out a region of quadsitrons which offers an optimal density, which translates into an optimal flow of energy within the realm of the electromagnetic energy spectrum. This need to continue to flow, will cause light, moving from point A to point B, to avoid a region in space where quadsitrons are packed together too tight as to not allow free movement of the quadsitrons in such a region of space. Light avoiding such a unique region of space, creates the phenomena of black holes dotting the universe.

CHAPTER 39

WORMHOLES EXPLAINED

WORMHOLES HAVE BEEN THEORIZED to connect two locations in the universe, a point A and a point B, in a different manner than drawing a straight line from point A to point B. Wormholes have been described as an imaginative means of shortening the distance of space travel, so as to facilitate characters of science fiction stories the means to cross vast distances of space in a shorter time than actually traveling the linear distance between point A and point B.

Einstein-Rosen bridges, named after Albert Einstein and Nathan Rosen, refer to wormholes between locations in space, modeled after Einstein's field equations.[123] Wormholes are suspected connections, such as a tunnel with two ends, each end of the tunnel in a separate point in spacetime. Wormholes are thought to be consistent with the general theory of relativity.[124] A wormhole is, in some instances, thought to act as a bridge between vast distances of even a billion or more light years, by physically transiting, in some cases a distance of just a few meters.

A wormhole might be likened to standing at the edge of a rock formation, two hundred meters above a ravine, facing a second rock edge one hundred meters distance away. If a bridge (wormhole) spanned the one hundred meters between the two rock edges, to cross the ravine, one could amble across the creaky bridge. If the bridge did not exit, in order to cross the ravine, one could choose to rappel two hundred meters down the rock face one had been standing upon, hike the hundred meters across the wooded ravine, and then climb the two hundred meters rock face on the opposite side of the ravine.

Herman Weyl (1885-1955), mathematician, theoretical physicist and philosopher, hypothesized a mass analysis of electromagnetic field energy speaking of one-dimensional tubes connecting points and spoke of spacetime geometry.[125,126] John Wheeler (1911-2008) coined the term 'wormhole' in 1957 with co-author Charles Minser (1932-).[127]

Physicists have described the possibility of a wormhole as being created by somehow folding space. Thus, a wormhole is possibly created by interpreting the universe as a flat pancake. Initially, the concept of a pancake sounds curiously inviting due to the Milky Way Galaxy having often been described as resembling a flat disc. Point A would be located on one end of the pancake and point B would be located on the opposing end of the flat, round pancake. The linear distance between point A and point B could be measured and would in general represent the actual distance of travel between point A and point B. Space theorists have thus described the existence of a worm hole as taking the pancake with point A and flipping half of the pancake over the top of the opposing end of the pancake, such that point B hovers over point A, thus in effect, dramatically shortening the linear distance between point A and point B.

This theory of flipping space in order to create shorter distances between two points in space admittedly sounds intriguing and even seems somewhat logical at first, until one realizes the universe cannot be likened to that of a flat pancake. The universe is the ultimate three-dimensional phenomena. When viewed from the perspective of our earth, the universe appears to be filled with an enormous amount of empty space, which indeed it is when one ponders the callosal distances between the stars. But when taken from the grand perspective of the Milky Way Galaxy as a whole, the billions of stars which comprise the galaxy represent substance, force and powerful gravitational fields. The universe cannot be frivolously flipped like a pancake, thus the linear distance from point A in a galaxy and point B is the 'true representative distance' between the two locations.

The vision of the great minds of theoretical physics has been that the universe is comprised of gravitational fields. Stars are surrounded by intense gravitational fields, and impose effects upon the planets in orbit around a star by means of gravity. This immense gravitational phenomenon extends across space, networking stars to each other. Gravity, thought to be a force, believed to act to ebb and flow throughout the universe, trapping light in black holes which are gravity wells. This matrix of gravity, causes an invisible, dynamic interplay between the stars comprising the universe.

If, the term 'gravity' was replaced by the term 'energy', then possibly the dynamics of the universe might come into clearer focus. The position taken in this text includes light is wave energy; magnetism is representative of trapping energy flow, moving molecules in some instances; gravity is representative of energy dense enough to manipulate the flow of quadsitrons, that is affecting the movement of the actual underlying fabric of the universe. Gravity speaks to the flow of energy in the universe. We do not appreciate the flow of energy toward the center of the Earth (i.e. gravity), because the surface of the planet and materials of the planet hinders a person from traveling to the core of the planet. To a SCUBA diver, participating in an open water dive, submerged in a lake or in the ocean, would experience a sense of the flow of energy about themselves by feeling the currents of water resisting or aiding the diver's capacity to swim, depending upon the direction the diver took. Gravity is likened to the flow of energy in the universe, intimately coupled with the quadsitron fabric of the universe; taking the example further, gravity is very much like water filling up a volume, such as a basin in some landscape on the planet's surface, with this water constantly flowing given inputs of additional water into this body of water from creeks, streams and rain, and water draining from the said body of water by rivers and evaporation. Gravity and water are constantly dynamic, both representative of the flow of unbridled energy and both representative of the flow of an underlying fundamental entity.

There is an alternative way to think of a wormhole. To this point, we have discussed energy, particularly light, as being a part of the electromagnetic energy spectrum. Energy, when accounted for in the context of the electromagnetic energy spectrum, travels through space in the form of a wave. The travel of energy as a wave is facilitated by the presence of quadsitrons, occupying three-dimensional space, acting as the transfer medium for wave energy. An analogy would be a wave rippling through water.

This text has also discussed the difference between the electromagnetic spectrum versus magnetic energy versus gravity being the density of the energy involved with each representation of energy. Gravity is more dense than magnetic energy, which is denser than the wave energy of the electromagnetic spectrum. As the density of energy increases, the characteristic wave formation of energy diminishes. It is well recognized that energy is segmented into irreducible quanta, therefore, how the quanta of energy traverse through space may be quite varied, taking on the shape of anywhere from a sinus rhythm wave to a straight-line wave. Since this is a discussion of three-dimensional space, the form energy might take would be three-dimensional ripple, which from a side view would look like a sinus wave, versus a column of energy with a varying central diameter, which from a side view would appear as a straight line.

Therefore, a wormhole coursing through space may be a long thin segment of extremely dense energy traversing through space. See Figure 166. The energy in a wormhole may be straight line energy, densely packed, traversing through space in one direction, without any wave motion. If there is no wave characteristic in the visible light range, the human eye would not be capable of detecting such a flow of energy. Alternative means of energy sensor would be required to detect the presence of a wormhole.

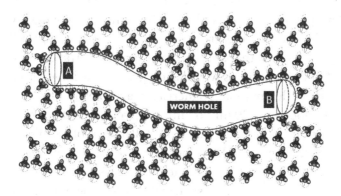

Figure 166

Wormhole in space is comprised of very dense
energy traveling through space.

The origin of a wormhole might be the result of an exploding star or other energy reservoir. See Figure 167. Given the intensity of the energy dispersed when a star ruptures, segments of very dense energy may be cast out into the three-dimensional space surrounding the exploding star. Such segments of very dense energy may then travel vast distances through the universe, possibly faster than the speed of light.

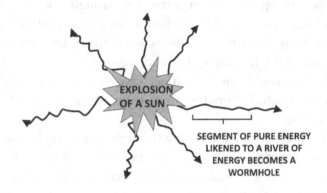

Figure 167

The origin of a wormhole may be the result of an exploding star,
ejecting dense segments of energy in varying directions.

Writers of science fiction novels have generally relied on one of three concepts to explain travel faster than the speed of light. Given the vastness of space and the great distance between stars, sci-fi authors have long recognized that if humans are limited to travel comparable to the speed of light, humans would age significantly long before arriving at any plausible destination. To increase travel beyond the speed of light or above a velocity of 720 million miles an hour, a sci-fi author generally relies on either space ships generating a means of propulsion that is greater than the speed of light, or the bending or warping of space to create the sci-fi version of wormholes as discussed earlier, or some form of jumping transport mechanism capable of leaping humans across vast distances in a very short period of time.

Authors of sci-fi also often alternatively use the concept of 'suspended animation' to explain how humans would travel through space without appreciably aging. The concept of suspended animation includes the idea of putting humans to sleep and at the same time appreciably slowing the body's metabolism down by utilizing a means of cryotherapy. It is generally expected that when such a space ship reached its destination the human occupants would be unfrozen and awoken. Presumably, years of being in a state of cryotherapy would halt the aging the process and the human occupants of such a spacecraft would rise from years of sleep, ready to engage the new world they had arrived at and would fit to venture forth and explore.

The concept of a wormhole may provide a means for humans to travel faster than the speed of light. Dense energy traveling in a straight line without the characteristic wave from seen with the electromagnetic energy spectrum, may travel faster than the speed of light, since such energy is not dependent upon quadsitrons to act as the transfer medium. Energy may be so dense in a worm hole that the density of the energy forces quadsitrons out of its path. An analogy would be a river flowing into a lake, where the stronger current of the river displaces the more still water residing in the lake. Energy in a worm hole moves from point A to point B. See Figure 168.

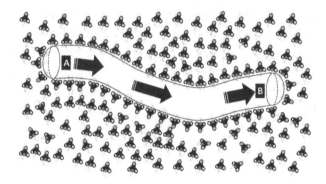

Figure 168

Wormhole comprised of dense energy traversing
from point A to point B in the universe.

If humans could construct a spaceship which could withstand the intense pressures generated inside the dense energy flow of a wormhole, and survive, human space travelers might be able to reach speeds faster than the speed of light utilizing wormholes as natural super highways. Like ancient explorers whom utilized rivers to make it easier to traverse uncharted territories, future space explorers may utilize wormholes to efficiently reach distant parts of the galaxy.

Wormholes representative of energy flow, may exist in one of two forms. One form of wormhole may be a directed energy flow, such as mentioned above when a star explodes. In this case, energy flows outward from the original source, then eventually dissipates some distance from origin, the energy redirecting, fading, and is some portions being absorbed by other components of the remaining universe. A second form of wormhole is likened to the channeling of energy seen with a bar magnet, as has been described in earlier chapters. A wormhole may exist as a flow of energy which loops upon itself, forming an endless closed circuit between various points in space. Thus, in some circumstances, a wormhole could be thought of as an energy loop.

If such energy loops exit, they would provide stable highways of travel between known points in the universe, which could be exploited for human interstellar travel. Such energy loops would likely be invisible to the human eye, given the density of energy and lack of radiant light in the visible light spectrum. Special energy detecting instruments would be required to locate such energy loops. Once inside an energy loop, the human traveler would be subject to being whisked along with the flow of energy, like a boat caught up in swift flowing rapids, and ending up where ever the loop of energy carried the space explorer. If the space explorer remained long enough within the energy loop, the traveler would complete the circuit and return to their original point of entrance into the loop.

CHAPTER 40

CONCISE THEORY OF QUADSITRON-ENERGY CONNECTIVITY

THE QUADSITRONS AND QUANTA are the basic building blocks of all matter. Right-handed and left-handed quadsitrons comprise the underlying fabric of the universe. The free quadsitrons act as an ocean extending in every direction reaching out to the very edges of the universe, suspending all matter within its fabric. Within the torrential pressures of the vortex of spiral galaxies, quadsitrons are fused together trapping quanta of energy to create neutrinos. Two quadsitrons, a right-handed and a left-handed, combine to trap one quanta. Some of the neutrinos are polar, with a positive and a negative pole. Some of the neutrinos are dark matter, with a neutral and an absolute zero pole.

Neutrinos combine to produce protons. Protons combine with the energy of an electron and a neutrino along with a quantity of quadsitrons which create a single layer of quadsitrons on the surface of the proton, to create a neutron. See Figure 169.

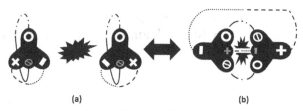

(a)   (b)

Figure 169

Quadsitron: (a) Right-handed and left-handed quadsitrons combine
with a quanta of energy to generate (b) a polar neutrino.

Polar neutrinos and dark matter neutrinos combine to generate protons. A proton is a stable spherical entity with positive poles of the surface neutrinos pointed outward. A single layer of quadsitrons along with a measure of energy adhere to the exterior of a proton to create a neutron. See Figure 170.

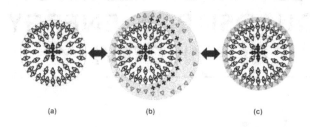

(a)   (b)   (c)

Figure 170

A single layer of quadsitrons and a measure of energy
cover the surface of a proton to produce a neutron.

Protons and neutrons combine to generate the nucleus of atoms. The subatomic composition of the nucleus of an atom generates orbitals around the nucleus. The orbitals are utilized by quanta of energy as pathways to orbit the nucleus of an atom. Orbitals are rings of quadsitrons formed out of the fabric of the universe. A segment of 463 rings of quadsitrons represents the boundaries of a quanta of energy circling the nucleus. See Figure 171. The number of protons and electrons dictate the type of atom. Hydrogen has one proton and one electron. Helium has two protons and two electrons. There are 118 recognized differing types of atoms. The neutrons present in the nucleus of an atom generally are similar in number to the number of protons of an atom. Often the number of neutrons vary, and these variants are referred to as isotopes of an element. For some elements, isotopes found in Nature are physically more stable than atoms with equal numbers of protons and neutrons.

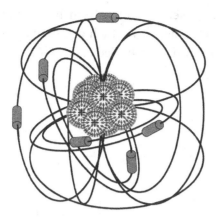

CARBON ATOM

Figure 171

Protons, neutrons and electrons combine to construct the architecture of atoms. The carbon atom depicted in this illustration has six protons, six neutrons and six electrons.

Atoms combine to form molecules. The glucose molecule which acts as the fuel source for many lifeforms, including humans, contains six carbons atoms, six oxygen atoms and twelve hydrogen atoms. See Figure 172. All organic lifeforms are the result of the combination of atoms, which combine to generate molecules to create the tissue structures and fluids, which form into bacteria, fungus, plants, animals and humans.

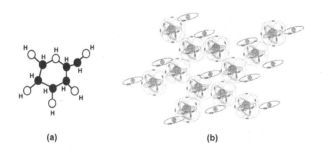

(a)   (b)

Figure 172

Schematic of the glucose molecule with six carbon atoms noted with a 'C', six oxygen atoms noted with a round circle and the position of twelve hydrogen atoms noted with the letter 'H'. (a) represents dot schematic of the glucose atom. (b) represents the molecular structures of carbon, oxygen and hydrogen comprising a glucose atom.

235

Carbon atoms, as well as atoms from numerous other elements, combine to construct molecules, which combine to make cells. Billons of cells comprise the human body. See Figure 173.

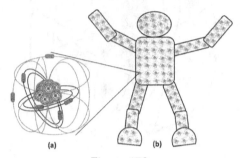

Figure 173

The human form is comprised of numerous carbon atoms and other elements organized together to form the tissues and fluids of the body.

The essence of the Theory of the Quadsitron-Energy Continuum is that every mass in the universe is connected to quadsitrons and energy. Quadsitrons combine to create neutrinos. Polar neutrinos and dark matter combine to construct protons and neutrons. Protons, neutrons and electrons combine to construct atoms. Atoms combine to create molecules. Molecules create organic structures such as the human body, as well as inorganic structures such as planets and stars. Stars and planets combine to make a galaxy. See Figure 174.

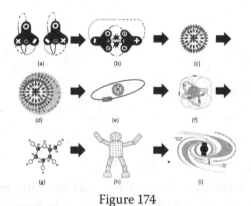

Figure 174

(a) Left and right handed quadsitrons; (b) polar neutrino; (c) proton; (d) Beta decay, (e) hydrogen atom, (f) carbon atom, (g) glucose molecule, (h) human, (i) Milky Way Galaxy.

The underlying thought, underpinning the entire context of this text is that quadsitrons are the physical fabric of the universe. Quadsitrons occupy and fill almost every space throughout the universe. See Figure 175. Though quadsitrons combine to construct protons, neutrons, electrons, atoms, molecules and all of the macro structs the human eye can see, a three-dimensional ocean of quadsitrons extends out in every direction a human can imagine. The end of the universe is the boundary where quadsitrons cease to exist.

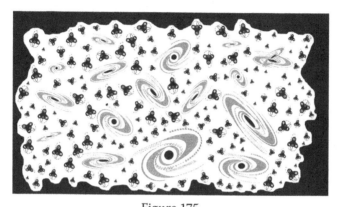

Figure 175

The Quadsitron Universe.

All in the universe is connected to the two underlying essential primary, nonreducible elements, that of quadsitrons and quanta of energy.

SECTION VI

ADVANCED THEORY

CHAPTER 41

THE 3 PHASES OF
WATER AND ENERGY

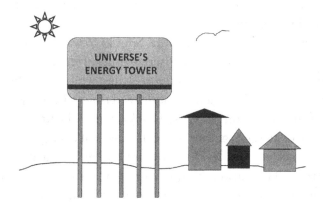

ENERGY IS LIKENED TO water. Where a body of water is typically a two-dimensional substance, energy is a three-dimensional substance. A body of water generally has air sitting over its surface, where the universe of energy is truly three-dimensional extending out in every conceivable direction. Where water, acted upon by gravity, generally exhibits a surface, such as the surface of water in a cup, or surface of a river or lake or ocean; energy in the universe, tends not to have a surface and exhibits characteristics and behavior in three dimensions.

Water is comprised of molecules of water. Each water molecule is comprised of one oxygen atom and two hydrogen atoms. Water molecules form into droplets. Water molecules can be trapped in a bottle or a cup. Droplets of water form into puddles of water. If rain falls from the sky, puddles may flow into creeks. Creeks of water and streams tend to follow the contour of the earth, seeking the lowest elevation. Streams can combine into rivers. Rivers of water can crash into each other and the river with the strongest current will direct the course of the river with the weaker current. As the density of the water increases, water begins to exert a force on the contour of the land, in some cases grossly reshaping the molecular face of the earth. As the density of water increases, a moving current can tumble boulders, plow through hardened dirt and given enough time, cut a path through solid rock, such as seen with the Grand Canyon. Rivers tend to flow into bodies of water such as ponds, lakes, seas and oceans. Lakes, seas and oceans tend to be comprised of varying currents of water, sometimes moving in differing directions. Tsunami's can generate one or more towering waves of water, which can crash into an island or the mainland with terrifyingly devastating effects.

Free energy is comprised of quanta. Similar to water molecules being discrete units of water, quanta are discrete units of energy. Quanta of free energy can be trapped in atoms and molecules, which then the energy is no longer considered free energy, but instead trapped energy circulating endlessly within loops, until released back out into the universe. Quanta of energy can coalesce together to create waves of energy, as the quanta move from one location in the universe to another using the fabric of quadsitrons. The various frequencies of wave energy make up the electromagnetic spectrum. As the density of quanta intensifies, the energy begins to flow and act like streams and rivers. Magnetic waves are energy that acts like a river, and when encountering other rivers of energy, the densest flow of energy will direct the movement of the weaker form of energy. Life on earth is protected by the magnetic field, which blocks the radiant energy from the sun from reaching the surface of the earth. Denser forms of energy, such as gravity, are capable of moving molecules and moves quadsitrons. The density of a stream of energy can reach that of a lightning bolt, where a lightning bolt is simply a very dense river of energy flowing from one location in the universe to another location in the universe.

In summary, energy occurs in four forms. See Figure 176. Energy exists as a wave form as described by the electromagnetic energy spectrum. The difference between energies represented in the electromagnetic spectrum is the frequency of the waves generated, or inversely the length of the wave generated; the slower the frequency the longer the wavelength. The energy of the electromagnetic spectrum represents a lower density of energy.

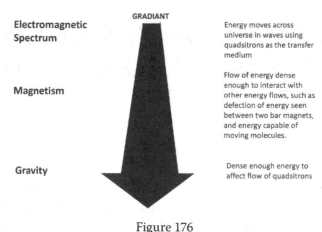

GRADIANT

Electromagnetic Spectrum

Energy moves across universe in waves using quadsitrons as the transfer medium

Magnetism

Flow of energy dense enough to interact with other energy flows, such as defection of energy seen between two bar magnets, and energy capable of moving molecules.

Gravity

Dense enough energy to affect flow of quadsitrons

Figure 176

The spectrum of density of energy.

Energy of the electromagnetic spectrum, moving as a wave, utilizes quadsitrons as a transfer medium to facilitate travel across the universe. Magnetism represents a density of energy greater than the electromagnetic spectrum and is sufficiently intense to affect the movement of molecules and atoms. Gravity is so dense in quanta of energy that this force is capable of moving or altering the flow of quadsitrons, the actual fabric of space. The fourth form of energy is that minute energy which is perpetually trapped inside the construct of a quadsitron and is responsible for the quadsitron's field lines; this, the smallest form of energy, calculates to be $0.854494198 * 10^{-20}$ joules.

CHAPTER 42

ADVANCING THE STUDY
OF ENERGY EQUATIONS

BESIDES THE EQUATION '1 + 1 = 2', Albert Einstein's equation $E = MC^2$ is the most famous and most recognized equation in the world.[128,129] The equation is short, simple, rolls off the tongue easily when spoken, fits conveniently on a tee-shirt, and conveys a relatively easy to understand concept. Einstein's infamous equation is symbolic of the distinct relation between the entities of 'energy' and 'mass'. Unfortunately, subconsciously, for most of us, the popular physics equation is more realistically: 'MODERN PHYSICS = WTF', where WTF may stand for your choice of 'What The F--K' or may take on the meaning: 'Wizardry, Theatrics, Fantasy'. The complexity of physics equations and the frustration brought on by physic authors' specific jargon and symbols, tends to cause the majority of the non-physics majors to be perplexed and distanced from the study of the subject. Physics is a part of everyone's daily life and the study of this essential subject matter should not be daunting, but instead, should be appreciated and embraced by all with intellectual curiosity...and it would be if the teaching of physics made better sense.

The equation $E = MC^2$ is descriptive of the energy of the universe being represented as equal to mass multiplied by the speed of light squared. Einstein published his work on this subject in 1905, one of a collection of four papers known as Annus Mirabilis (Miracle Year) papers.[130] People all over the world have recognized this distinctive equation as the universal symbol of physics, yet a realistic concrete understanding of what the equation actually represents remains elusive.

Energy is generally identified as joules. Mass is generally identified as some measure of kilograms. The speed of light is generally represented as meters per second, or miles per hour, or miles per second. Joules can be reinterpreted as kilograms times meters per second.

In 1946, Conférence Générale des Poids et Mesures (CGPM) Resolution 2 standardized the unit of force in the MKS system of units to be the amount needed to accelerate 1 kilogram of mass at the rate of 1 meter per second squared. In 1948, the 9th CGPM Resolution 7 adopted the name 'newton' for this force.[131] The 'newton' thus became the standard unit of force in the International System of Units.[132]

James Prescott Joule (1818-1889), studied heat and its relationship to mechanical work. The 'joule', a unit of measurement of energy, is named after him. A single 'joule' is equal to the amount of energy or the amount of work of a force of one 'newton' transferred to an object along the path of travel of the object through one meter of distance. A joule can be described as one newton-meter. The joule is commonly also described as kg per meter^2 divided by seconds^2.[133]

Max Planck created the energy equation $E = h * v$, where E is energy measured in joules. In Planck's equation 'v' represents frequency expressed in Hertz and 'h' represents the constant Planck himself derived.[134],[51] Planck's constant is 6.626×10^{-24} joules * second.[135] The equation represents energy in the form of a sine wave, which across the various frequencies, compiled together, is the electromagnetic spectrum. Thus, since frequency is equal to the speed of light divided by wavelength, Planck's equation can also be represented as $E = h * c / \lambda$, where 'h' is Planck's constant, 'c' is the speed of light and 'λ' represents wavelength. Planck's equation is meant to represent energy of one complete sine wave at a given frequency, which has been referred to as a packet of energy, or as 'quanta' of energy or as a 'photon' of energy.

Planck's equation depicts Planck's constant 'h' and frequency 'v' represent the amount of energy in joules for one complete wavelength of an energy wave at a given frequency. The higher the energy, the more cycles of sine waves for a given period of time, which is represented as frequency described in Hertz which is cycles per second. Therefore, utilizing Planck's equation, given a specific frequency of an energy wave in the electromagnetic spectrum, the amount of energy associated with the wave at the specific frequency can be determined.

Where Planck derived his equation $E = hv$ as a means of describing the energy in a photon (single wave) of the electromagnetic spectrum; Albert Einstein's equation for energy $E = MC^2$ represents energy confined within an object derived from the mass of the object.

For the equation $E = MC^2$, energy is represented again in units of joules. The mass is in units of kilograms. The speed of light is in meters per second squared. Both equations, $E = h^* v$ and $E = MC^2$, produce results in units of 'joules', which is also represented as kg-m^2/sec^2. Einstein's work established the correlation of 'mass' as it relates to 'energy'.

The second part of Einstein's energy equation is 'mass'. The subject of mass becomes more curious as the study of the subject deepens. Mass is often thought to be synonymous with weight. Actually, there is a significant disconnect between the properties of mass and weight. Both systems of measurement represent the same thing when resting on earth. As we know, when an object is relocated into space or placed on the Moon, the weight of the object changes. Inherently, since mass is representative of energy of the object, the mass should not change where ever in the universe the object resides.

If an object has a weight of 12 kilograms and a mass of 12 kilograms on the earth, then on the moon the weight will drop to 1/6 proportion of that on the earth or be approximately 2 kilograms, the mass of the same object remains 12 kilograms. So, the difference between mass and weight is that mass is somehow tied to energy contained in the object and is always constant, while weight is somehow tied to the gravity at the location the weight of the object is being measured.

The concept of mass tends to erode when mass is considered to be measured by a three-point balance scale, where the mass of an object to be measured is compared to a known quantity. Problem is this seems to only work on the surface of the earth. Taking a three-point balance scale and conducting the same measurement in zero gravity would suggest all objects have the same mass, that is zero.

The other confusing thing about 'mass' as represented in the equation $E = MC^2$ is that since the speed of light is supposed to be a constant, mass is the only variable on the right-hand side of the equal sign. Mass represents the energy in question if C^2 is a constant. Mass must therefore represent the energy confined in molecules and atoms that comprise whatever substance is being evaluated by this equation.

A 'mole' of a substance is used to represent a unit number of elements in a substance. One mole of a substance is considered to be 6.02214076 * 10^{23} particles of that substance.[136] Therefore, a mole of water as a solid has the same number of water molecules as a mole of water as a liquid and the same number of water molecules as a mole of water vapor. The space occupied by the mole of water in solid form will be different than a mole of water in a liquid form and in a gaseous form, but again the same number of water molecules is present in each of the three states of matter if one mole is used as the unit of measure.

The molar mass of any atom or molecule is equal to the atomic weight of the atoms. The molar mass of ice is 18.01528 grams. In addition, one mole of water as a solid, equals the weight of one mole of water as a liquid which equals one mole of water as a gas. We know that atoms are storehouses of energy. If mass is representative of energy, but mass of equal amounts of a substance do not change from solid to liquid to gas, then the question which arises is 'how does mass account for energy put into atoms, causing electrons to orbit at more distant orbits from the nucleus'.

The equation $E = MC^2$ was derived in 1905, where the 'C' representing light speed was considered a constant. Later, Einstein subscribed to the concept that the velocity of light could be manipulated by natural forces. Einstein taught that a black hole was a celestial phenomenon that trapped visible light. The quantity of gravity of a black hole was so intense that light could not escape such a phenomenon. Trapping light within a black hole would logically translate into the velocity of light had been altered, if not decreased to zero. There is the possibility that light is continuously circulating, maintaining the velocity of the speed of light, within the confines of the black hole, but then there is some other force at work which has yet to be explained.

Einstein received the Nobel Prize in Physics for accurately predicting that light from a distant star, passing by a gravitational body such as our solar system's sun, would bend toward the gravitational body. This was proven in 1919. Einstein was awarded the Nobel Prize in 1921. Given again that a gravitational body could influence the pathway of light, one must consider that the velocity of light is not held constant if the forward path can be manipulated.

$E = MC^2$ does not appear to have sufficient flexibility to be used in venues other than in a static setting, where the object in question is in a solid form and the object is resting on the surface of the earth.

Max Planck's equation $E = h^* v$ represents energy in a single wave of energy, which is referred to as a photon. Max Planck's equation serves the purpose of demonstrating that there exists energy in the waves of the electromagnetic spectrum, and that the energy of a wave at a certain frequency can be calculated. Planck's equation is based on a photon being the measure of a single wave. Planck's photon has been synonymous with a quanta of energy. Planck designated that a photon was the elemental unit of wave energy, which is a single wave. But as demonstrated by utilizing Planck's equation, each photon of a different frequency has a different energy level. Thus, Planck's equation measures the energy of a wave, but does not provide true representation of the energy of a single quanta of energy regarding the definition of a quanta being the most elemental unit of energy in the universe. Some unit of energy must make up the waves of energy per frequency in the electromagnetic spectrum.

Both equations $E = h^* v$ and $E = MC^2$ have limited utility, confined generally to use on the surface of the Earth and limited to the environment and the gravitational state of the object being measured.

If humans wish to venture beyond the boundaries of the earth, then the concept of energy must be expanded to meet the needs of space adventurers.

As the captain of an interstellar spacecraft, the equation $E = MC^2$ falls short of the requirements of a space ship venturing beyond the safe haven of Earth. The spacecraft is in essence in a weightless environment, thus mass in kilograms becomes a relative quantity and the spacecraft is traveling near or at the speed of light, where meters per second is a nonconsequential means of measurement. If the spaceship's captain and crew has ventured too close to a star which is in the process of exploding, the captain is going to be most interested in the resources he has available to strengthen the energy shields which protect the ship. The density of the energy shields protecting an interstellar space ship will be gauged in number of quanta of energy available to fortify the shields. A captain trying to protect his ship and crew from the radiation output by an exploding star will be most interested in how many quanta of energy he has at his disposal to strengthen his protective energy shields surrounding his ship.

To derive such an equation, one recalls that energy is measured in joules. Derived earlier in this text was the energy of one neutrino, considered the energy of a 'quanta'. This energy quanta measured in Ev/c^2 is not to be confused with Planck's definition of a 'quanta', which is measured in joules. The energy equation using quanta measured as Ev/c^2 is derived as described below:

$$E = (\text{Energy 1 quanta in Ev/}C^2) * (\text{conversion Ev to joules}) * (\text{number of quanta}) * C^2$$

$$E = 0.1066 \text{ Ev/}C^2 * 1.6021766208 \times 10^{-19} \text{ J/Ev} * Q * C^2$$

$$E = 0.17079202777728 \times 10^{-19} \text{ J/ } C^2 * Q * C^2$$

$$E = 1.7 \times 10^{-20} \text{ J/ } C^2 * Q * C^2$$

$$E = 1.7 \times 10^{-20} \text{ Q joules (energy at rest)}$$

Thus, energy as 'E' measured in joules, is equal to 1.7×10^{-20} times then number of quanta of energy represented as 'Q'.

This equation allows a starship captain to comprehend how much energy is available to strengthen the protective shields of the ship as measured in number of quanta held in reserve.

This equation also identifies that for one quanta the amount of energy is $1.7 * 10^{-20}$ joules.

This is also the amount of energy trapped inside a neutrino, therefore a starship captain may measure her/his available fuel as $E = 1.7 * 10^{-20}$ joules Q or as $1.7 * 10^{-20}$ joules times the number of neutrinos stored in the ship's fuel tanks.

CHAPTER 43

ADDING TO THE QUANTA ENERGY EQUATION

Energy = [MC^2] + [6.4x10^{-20} * p] + [15.3 x10^{-4} * p
+ 15.3 x 10^{-4} * n] + [excitable energy]

ALBERT EINSTEIN'S ICONIC ENERGY equation E = MC^2 is an energy equation which related mass of an object to the energy within the object. Mass is arrived at by measuring the mass of an object against the mass of a known object using a triple beam balance. The triple beam balance is named after the mechanism which details three beams which allows a known mass to be moved along a beam. A vivid example would be the weighing scale used in old doctor's offices, which incorporated a series of weights that slid on a bar, to arrive at a measure of a patient's weight.

In the study of physics, the argument is made that weight and mass are different. On the surface of the earth, weight and mass are virtually considered equal. Placing an object on the surface of the moon, where the gravity is one-sixth that of the earth, an object's weight would be one-sixth of the weight on earth, but the mass would be considered the same on earth or on the moon.

If one took a wood log and measured the weight, the log would have a certain weight given the size and water content. If one transported the log to the moon, although nothing would change regarding the features of the log of wood, the weight would adjust to the moon's gravity and the log would weigh one-sixth of the weight the object had resting on the earth's surface. Further, if one took the log of wood in a space ship to a location beyond the gravity field of any moon or planet, the weight of wood in a weightless environment would be zero. Yet, whether the log of wood where located on the surface of the planet earth, or located on the surface of the moon or in a vessel in space, given the proper conditions, the log of wood could be burned in a similar manner by igniting the log on fire. Regardless of the gravity field the log of wood resided in, and thus regardless of the weight of the log of wood, the log would retain its inherent potential energy, which could undergo combustion resulting in fire. Hypothetically, if provided similar conditions such as surrounding oxygen and nitrogen content, air pressure, humidity, temperature, and radiation levels, the log should burn similarly in all locations.

If one took the wood log and measured the mass of the log on a triple-beam balance on the earth's surface, one would arrive at a weight and mass. Alternatively, if one randomly selected wooden log from a forest on the earth's surface, took the log up into space and then measured the mass of the log on a triple beam balance in a weightless environment, the known weight would be zero, the weight of the scale would be zero, and the mass of the log would be zero given all objects including the vehicle would be weightless. This suggests that mass is dependent upon measuring the mass of an object on the surface of the planet. Yet, the capability to burn the log of wood, is retained whether the log is resting on the earth or whether the log is in space.

Mass of an object, must somehow be connected to energy in a manner different than measuring the mass of an object on earth.

The absolute total energy of an object is analyzed in the context of an object passing through the vortex of a galaxy, and the extreme forces of the vortex rip the object completely apart, disassembling the object down to the individual quadsitrons and energy which comprise the object.

There are several ways a solid object contains energy. The first form of energy is contained in the molecular bonds comprising the object. The second form of energy is contained in the electrons orbiting atomic nuclei of the atoms of the object in the nonexcitable state. The third form of energy is locked in the neutrinos comprising the protons and neutrons of the atoms of the object. The fourth manner energy is present in an object, is the amount of energy stored in the excitable electrons of an object. An excitable electron refers to an electron which achieve a more distant orbit from the nucleus of an atom, due to the atom having absorbed energy from an external source. The equation $E = MC^2$ is sufficient on the earth's surface, but is not fully representative of the total energy contained in an object entering the vortex of a galaxy.

Energy = [molecular bonds] + [electrons per atom] + [protons + neutrons] + [excitability]

Energy = E1 + E2 + E3 + E4

The energy contained in the molecular bonds is assumed to be represented by $E = MC^2$.

$E1 = MC^2$

The number of electrons present in an atom generally correlates with the number of protons present in the nucleus of an atom. The energy contained in the electrons of each atom is generally recognized by counting the number of protons in the atoms comprising the object.

An estimate of the energy contained in each electron is provided by the energy released in the Beta decay of bismuth-210 to polonium-210, which is 0.4 MeV. The products of this beta decay include a proton, an electron and an antineutrino. The energy related to the neutron decaying to a proton is 1.16 MeV, which is the sum of the energy of the electron, the antineutrino and the recoiling of the nuclide.

E2 = quanta of energy * number of protons

E2 = 0.4 MeV * protons

E2 = 0.4 MeV * protons * 1.6021766208 J/eV

$E2 = 0.64087064832 \times 10^{-19}$ J * protons

The energy confined by the neutrinos comprising each proton and neutron is represented by the number of neutrinos comprising a proton multiplied by the number of protons in an object, added to the number of neutrinos comprising a neutron multiplied by the number of neutrons in an object.

E3 = quanta of energy * number of neutrinos in a proton * number of protons + (quanta of energy * number of neutrinos in a proton + energy of beta decay) * number of neutrons

E3 = 0.1066 eV/C^2 * 8.7963 x 10^9 * protons + (0.1066 eV/C^2 * 8.7963 x 10^9 * + 1.16 MeV) * number of neutrons

Conversion eV/C^2 to eV =

0.1066 eV/C^2 * 8.98755179 x 10^{16} m^2/sec^2 = 0.958073020814 x 10^{16} eV

= 9.58073020814 x 10^9 MeV

E3 = 9.58073020814 x 10^9 MeV * protons + (9.58073020814 x 10^9 MeV + 1.16 MeV) * neutrons

E3 = 9.58073020814 x 10^9 MeV * protons + (9580730208.14 x MeV + 1.16 MeV) * neutrons

E3 = 9.58073020814 x 10^9 MeV * protons + 9580730209.3 x MeV * neutrons

Conversion of eV to joules:

E3 = (9.58073020814 x 10^9 MeV * protons + 9.5807302093 x 10^9 MeV * neutrons) * 1.6021766208 x 10^{-19} J/eV

E3 = 15.35002194967 x 10^{-4} J * protons + 15.35002195153 x 10^{-4} J * neutrons

E4 represents excitable energy. E4 = number of quanta added to the excitable electrons in the atoms of the object. E4 is a dynamic measure in time. This is related to the energy absorbed by an object. The energy in the form of E4 can be thought of as heat absorbed by an object from a heat source; in such a situation the mass of the object does not change since the number of protons, neutrons and electrons stay the same, but the atoms comprising the object contain additional energy with electrons orbiting in higher orbits. E4 may fluctuate over time, increasing or decreasing. An object may release energy stored in the object as excited electrons fall back to orbits closer to the nucleus of the atom. E4 may only be appreciated by knowing how much energy has been added to the object.

An example of excitable energy would be liquid water turning to water vapor. A cup of liquid water resting in a pan consists of a specific number of water molecules. If the pan is heated up to the boiling point of water, the individual water molecules absorb the heat energy. The excited water molecules become airborne as they transform into water vapor or steam. In a science laboratory, the entire cup of water could be confined. The cup of liquid water would have the same mass as a cup of water vapor, though the water vapor under room pressure would occupy a larger volume than the cup of liquid water. The cup of water vapor would contain more energy than the cup of liquid water per kilogram weight.

Summing up the energies

Energy = [molecular bonds] + [electrons per atom] + [protons + neutrons] + [excitability]

Energy = E1 + E2 + E3 + E4

Energy = $[MC^2]$ + $[0.64087064832 \times 10^{-19}$ J * protons] + $[15.35002194967 \times 10^{-4}$ J * protons + $15.35002195153 \times 10^{-4}$ J * neutrons] + [stored excitable energy]

Energy = $[MC^2]$ + $[6.4 \times 10^{-20}$ * p] + $[15.3 \times 10^{-4}$ * p + 15.3×10^{-4} * n] + [excitable energy]

CHAPTER 44

COMPLETING THE QUANTA ENERGY EQUATION: THE ADDITION OF MOMENTUM

$$E = [MC^2] + [6.4 \times 10^{-20} \, {}^*p] + [15 \times 10^{-4} \, {}^*p + 15 \times 10^{-4} \, {}^*n] + [\text{excitable energy}] + \Sigma^9 \, M\phi^*C$$

ALBERT EINSTEIN'S EQUATION $E = MC^2$ is generally applicable to a solid object present on the earth at a state of rest. For humans living on earth, the perspective of $E = MC^2$ is suitable for the general needs of physics analysis and interpretation. But venturing beyond the scope of the surface of the planet requires a broader application of such an equation and therefore, an expansion of the energy equation.

If one invokes Einstein's Theory of Relativity, the perspective of the solid object changes for a viewer whom is not residing on the surface of the earth, but whom is at a distant location in space. The Theory of Relativity dictates that the viewer in space will witness a behavior that is different than the viewer residing on the planet's surface. The view of actions on earth, from a distant vantage point in space, is quite different than what is witnessed by a person participating in an action on earth.

Weight of an object relies on gravity, and often, right or wrong, weight and mass are interpreted as the same entity. Therefore, Einstein's equation is limited in two manners. One, change the gravitational field and mass changes. Second, if a black hole actually influences light, or any other gravitational body can influence light, then the speed of light is not a constant. Therefore, Einstein's equation $E = MC^2$ may only be applicable to the single-frames references on earth. What of reference planes beyond the surface of Earth?

When quantifying energy with regards to a particular object, one needs to take into account the concepts of pure trapped energy and the concept of potential energy. Einstein was correct in his Theory of Relativity. Concepts in physics are relative to the frame of reference the object resides in as well as the frames of reference surrounding the object.

For humans to venture forth from the earth and moon, and become mobile in deep space, a new concept of energy needs to be considered. $E = MC^2$ is representative of a static object in a static single-frame reference. The universe on the other hand is dynamic regarding numerous dimensions. Expansion of the energy equation is a necessity for space travel, if one were to venture from one star in the universe to a second star in the universe.

Let us take an example where one becomes lost in the woods with only a day pack to survive the night. The reader, you, are cast in this short narrative in the role of the unexpected camper. Alone, miles off trail standing at the side of a small river, no cell phone service, with the sun slowly sinking on the horizon, you decide to set up camp for the night. You decide to construct a shelter on a cliff next to the river. This serene appearing river feeds a waterfall a hundred yards downstream. The waterfall drops a hundred feet down into a plush canyon, where the river continues to make its way through the dense forest disappearing in the horizon. It will be dark soon, so building a fire becomes essential to both keep warm and to stave off potential predators during the night.

You venture into the thorny woods to hunt for firewood. You gather twigs to start the fire, and return to the campsite. You know you are capable of starting a fire with the Firestarter in your daypack, but to sustain the campfire through the night will require a number of sizeable logs of wood. You retrieve a saw from the daypack and venture back into the forest. You come upon a downed tree with a trunk about the width of a clenched fist. The appearance of the exterior bark suggests the tree has been dead for a sufficient amount of time to make the wood suitable for burning. You start with the thinner sections of the downed tree and begin cutting the trunk of the tree into two-foot sections. You return to camp with an arm full of cut logs. You toss the logs onto the ground next to the pile to twigs you had gathered earlier.

The equation $E = MC^2$ is a what is thought to be a mathematical equation to predict the amount of energy carried by each log to support the fire you plan to build. Mass of each log in the pile you gathered multiplied by the speed of light squared should predict how much energy you, the lost camper, has available to burn in the fire. By measuring the mass of each log, then determining burn time of each log given the time from sunset to dawn, given the median air temperature throughout the night, wind speed, elevation above sea water which is related to oxygen density in the air, residual moisture in the logs, a mathematics savvy camper could determine, before night fall, if the amount of wood gathered from the forest would be sufficient to keep a fire going all night long.

Believing there is a sufficient number of logs to sustain a fire, the camper [Notice the frame of reference has been changed] builds a fire on the plateau overlooking the waterfall. After successfully igniting the twigs, and carefully piling the smaller logs onto the initial kindling, a raging camp fire burns in the makeshift fire pit.

As the camper sits next to the crackling fire, a gentle breeze lifts the smell of burning pine up such that the camper can appreciate the sweet scent of the burning wood. Thirty minutes pass and another log is required to maintain the fire. The camper reaches over to the pile of wood logs and retrieves one.

Holding one of the logs firmly in hand, the camper suddenly has an idea. Lost in the woods, with no electronic means of signaling for help, the camper conjures up an idea to write their name on the log of wood with a message requesting assistance, drop the log into the river coursing through the canyon below. Like a message in a bottle, possibly the log would reach someone located downstream who would recognize the call for help and send a search party.

The camper strips the bark from the piece of wood and carves a message on the log. Standing at the edge of the cliff, the camper peers out over miles of woodland, with nothing human in sight. Gazing down the jagged rock face of the cliff, a mist plumes upward as a river of water fifty feet wide rushes to tumble a hundred and fifty feet down to a small reservoir nestled in the valley below. The river continues for a half mile, cutting its way through the valley floor, until it makes a sharp bend to the right and disappears view behind the trees.

The camper tosses the log over the cliff aiming for the reservoir below. At this point, $E = MC^2$ becomes old school in the vastness of the universe. Previously, it was established that $E = MC^2$ represented the energy trapped inside the wooden log if the log were burned. Now, as the log tumbles toward the reservoir below, the log is increasing velocity as it falls and it is twisting. During the act of falling through the air, the energy the log represents has increased above and beyond $E = MC^2$ due to the accelerating velocity of the log and due to the momentum related to the twisting motion the log is participating in as it drops toward the reservoir below. As the log falls there is at least $E = MC^2$ + acceleration multiplied by distance + momentum.

The camper described in the above illustration, exists on the surface of the planet earth. To the camper's perception, all of the objects that surround the camper are essentially stationary. As described by Albert Einstein's Theory of Relativity, the camper views the objects that share the camper's frame of reference as being stationary with respect to the camper. What the camper could consider, utilizing the essence of the Theory of Relativity, are the various unseen momentums acting on the log as the log falls from the cliff where the camper stands down to the reservoir below. If the camper stays on earth, compliant with the Theory of Relativity, there is no need to account for the unseen momentums and speeds of travel. If the camper aspires to someday become the captain of an interstellar spacecraft and travel the universe, the camper will need to take into account the unseen momentums which act on any object in the universe when such an object travels from one frame of reference in the universe to another frame of reference in the universe.

As the log falls from the hands of the camper toward the reservoir below, the forces acting on the log, as described previously, include acceleration and a twisting motion which creates momentum. This is the view from the camper's perception. The view of the same event from the frame of reference of a witness standing on an observation deck in a spacecraft located outside the boundaries of the Milky Way Galaxy would see at least the following additions to the energy signature of the log:

1. the log's inherent energy, including momentum created by movement of the log on the surface of the earth;
2. the log moving in a circular orbit around the sun at 67,000 miles an hour;
3. the log twisting in space as the earth rotates on its axis;

4. a momentum created by the wobble in the earth due to variations in the orbit of the planet as the planet circles the sun;

5. momentum created by the solar system as a unit circulating about the center of the Milky Way Galaxy as part of the Orion Spur; See Figure 177;

6. the momentum created as the solar system as a unit tumbles through space;

7. the momentum created by the Milky Way Galaxy as the collection of stars moves as a unit through the universe;

8. the momentum generated by the universe as a unit expands/contracts as the collection of stars move through space;

9. the momentum created by the Milky Way Galaxy as the collective of stars tumble through space.

The viewer, standing on the observation deck of a spacecraft located outside the boundaries of the Milky Way Galaxy, spying the log falling from the cliff would calculate the energy of the falling log to be $E = MC^2 + \Sigma^9$ Momentum * C, where at least nine differing momentums of energy are acting on the log as the log falls from the hand of the camper toward the reservoir at the bottom of the waterfall. Momentum must be multiplied by the speed of light to arrive at joules for the energy equation.

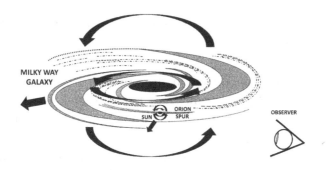

Figure 177

There are at least nine differing momentums acting on any object carried by the Earth. The Sun's solar system is tumbling. The Milky Way Galaxy as a whole is tumbling through space.

Now, alternatively, just before the camper lets the log fall from their hand, the camper decides to board an intergalactic spacecraft. The camper carries the wooden log on the journey to the center of the galaxy. The spacecraft transports the camper from the earth to the spiraling vortex referred to as the Milky Way Galaxy's black hole. At the center of the galaxy the camper tosses the wooden log into the galaxy's black hole. The forces of gravity are so strong within the boundaries of the black hole, that not only are molecules are ripped apart, then all of the atoms comprising the wooden log are ripped apart, but the individual protons, neutrons and electrons are ripped apart down to the two essential elements, and the wooden log is reduced to that of energy and quadsitrons.

Therefore, to satisfy Einstein's Theory of Relativity and the view a person would have located in space some distance from the earth, the energy equation would need to have momentum added. Momentum is the quantity of motion of a nonstationary body measured by mass multiplied by velocity. The momentum (Mϕ) for the situation as it applies to this discussion would need to be calculated in at least nine different frames of reference.

'p' is the number of protons in an object; 'n' is the number of neutrons in the object.

$$E = [MC^2] + [6.4 \times 10^{-20} \, {}^*p] + [15 \times 10^{-4} \, {}^*p + 15 \times 10^{-4} \, {}^*n] + [\text{excitable energy}] + \text{Momentum}$$

Momentum is measured in mass times velocity. The units of momentum are kilograms times meters per second. To fit momentum into an energy equation, where the units are joules, the momentum requires modification. Joules are measured in kilograms times meter^2 per second^2. In order to achieve the joules unit of measure, momentum (Σ^9 Mϕ) is multiplied by speed of light (C), the speed of light being measured in meters per second.

Taking the modified momentum into account, the final equation for energy becomes:

$$E = [MC^2] + [6.4 \times 10^{-20} \, {}^*p] + [15 \times 10^{-4} \, {}^*p + 15 \times 10^{-4} \, {}^*n] + [\text{excitable energy}] + \Sigma^9 \, M\phi \, {}^* \, C$$

This energy equation includes mass measured as weight on earth, energy associated with the electrons in orbit around the atoms of an object, energy stored in the protons and neutrons of an object, the excitable energy contained in the object and the momentum of the object being a part of the overall universe.

If $E = MC^2$ is replaced by the quanta energy equation 1.7×10^{-20} Q joules, which is independent of mass as measured on earth, then the energy equation converts to the following:

$$E = [1.7 \times 10^{-20} \, {}^*Q] + [6.4 \times 10^{-20} \, {}^*p] + [15 \times 10^{-4} \, {}^*p + 15 \times 10^{-4} \, {}^*n] + [\text{excitable energy}] + \Sigma^9 \, M\phi^*C$$

But in this case $[6.4 \times 10^{-20} \, {}^*p] + [15 \times 10^{-4} \, {}^*p + 15 \times 10^{-4} \, {}^*n] + [\text{excitable energy}]$ have already been captured and accounted for in the quanta equation $1.7 \times 10^{-20} \, {}^* $ Q. Thus, the completed equation becomes:

$$E = 1.7 \times 10^{-20} \, {}^* Q + \Sigma^9 \, M\phi \, {}^* C \quad [\text{Representative of energy of an object in motion}]$$

as apposed to: $E = 1.7 \times 10^{-20} \, {}^* Q$ [Representative of energy of an object at absolute rest]

Again, energy of an object taking in the majority of the frames of reference in the universe:

$$E = 1.7 \times 10^{-20} \, {}^* Q + \Sigma^9 \, M\phi \, {}^* C$$

This energy equation includes quanta energy and the momentum of the object as it resides as an integral part of the overall motion of the universe. The above energy equation would be considered a dynamic energy equation. The absolute speed of light as mentioned above is referenced to a stable flow of quadsitrons in the universe. In the case of a stable flow of quadsitrons in the universe, the speed of light is likely to, in fact, be constant, contrary to the current belief that gravity can deflect light and black holes are capable of trapping light. Light, like other energies comprising the electromagnetic spectrum, likely use the transfer medium of quadsitrons in a relatively uniform manner throughout much of the universe. Alternately, violent or catastrophic flows of quadsitrons, such as associated with the vortex of a black hole of a galaxy or explosion of a star, may re-direct the flow of light and affect the speed of light in certain situations.

CHAPTER 45

ABSOLUTE TIME, THE ONLY TRUE CONSTANT IN THE UNIVERSE

ALBERT EINSTEIN TAUGHT THAT gravity, velocity of an object and time were all related. Einstein described a universe where celestial bodies with large mass warped gravity and seemingly warped time. Einstein further taught that the faster an object traveled, the slower time would pass for such an object.[137] Two key concepts that have been dramatized; the first being: A clock residing on the surface of planet earth moves slower than a clock positioned further away from the center of the planet such as a clock on a mountain top or a clock in orbit around the earth. This phenomenon of time discrepancy related to distance from the center of the planet has been proven, and the GPS system which assists in navigation and is dependent upon data gathered by satellites in orbit around the earth, must account for this phenomenon to maintain accuracy for the end users of the technology whom reside of the planet's surface. The second concept is a clock resting on the surface of the earth records time faster than a clock which is moving at high speed, such as a clock onboard a space ship in orbit around a planet or traveling from the earth to the Moon or another planet. Again, this second concept has been proven to be true by matching time recorded by a clock resting on earth and a clock in orbit around the earth.

First, it must be understood that what one might consider a clock sitting on a mantle above a fireplace to be stationary because this is what human senses tell us, but in fact such an object is not actually at rest. Planet earth and everything on the earth is hurdling through space with a forward velocity of 67,000 miles per hour. The earth is also rotating as the planet orbits the sun. The entire solar system, of which the earth is the third planet from the sun, is part of the Orion Spur, which is circling the center of the Milky Way galaxy with a single revolution taking approximately 120 million years. In the scheme of the universe, nothing on earth is truly sitting at rest. What sits on earth, seemingly at rest, is only a baseline of motion, to be used to compare to other objects which are exhibiting alternate forms of motion.

So, the 'phenomena of time' is a commodity, which can be altered as Einstein predicted and has been proven,...or is it? Is Einstein's prediction simply a statement which captures and stimulates the human imagination, while at the same time seemingly satisfying human perception within the scope of the limits of our senses. The essence of what time consists of has not yet been explained.

Time is a human fault. Time has no meaning to the matter or energy of the universe. The timing of eruptions, collisions, implosions, explosions means nothing to inorganic matter and more than likely are inevitable and cyclic. The measure of time is a means for humans to mark off segments of their own existence, whether it be years, days, hours, minutes, seconds, micro-seconds or other forms of dividing up the human existence. Humans use the measurement of time to conduct and keep track of various processes associated with the human existence. The concept of time as humans measure time, is not an absolute, but instead very much dependent upon the means of measuring time.

The human measure of time being set in motion by the rotation of the earth creating day/night cycles, and the orbit of the moon creating a monthly cycle. Later, the study of the constellations embedded in the night sky assisted in understanding the timing of seasons. Creating a yearly calendar based on phases of the moon and star patterns assisted farmers with understanding growing seasons, optimizing the planting of seed and harvesting of crops.

In a universe where no luminous aether exists, a clock can be described as a device which is not internally affected by movement through the space or location in space. Thus, in a universe where no luminous aether exists, the measurement of time by a clock, can in fact be construed as being a measure of time without additional effect on the measure of time other than change in position of two clocks or difference in speed of two differing clocks.

The concept of recording time dramatically changes when the concept of a luminous aether is introduced into time measurement exercises. Taking into account the effect of the luminous aether, one must account for the fact that any measurement device, including an atomic clock, will be affected by the ebb and flow of the luminous aether, which infiltrates and surrounds all matter, and plays an active role in all matter at the sub^2 level. A clock located closer to the earth center will operate in an environment where the density of the quadsitrons is higher than a clock located further distant from the core of the planet. A higher density of quadsitrons closer to the center of the planet will make the internal workings of a clock, either mechanical or atomic, work slower, than a similar clock located further away from the core of the planet. Similarly, a clock moving at velocity through space will encounter a higher density of quadsitrons bombarding the surface of the protons and neutrons comprising the matter of the device, then a clock sitting at rest. The faster the clock moves through space and becomes bombarded by an increasing number of quadsitrons, the slower the clock will mark off time. It would be similar to an athlete running a mile on an open track versus an athlete running a mile through waist deep body of water. Both runners would accomplish the mile run, but the density of the water versus density of air would cause the athlete running through water to run slower. The internal workings of a clock, whether it be mechanical or atomic, moving through the fabric of space, will operate slower and record time slower, than a clock at rest.

Albert Einstein was correct regarding his predictions of the phenomenon of the warping of time in relation to two clocks behaving differently in two separate locations in space. The basis of this warping of time though is more than likely a phenomenon of limitation of the measurement of time and the effect of the fabric of the universe on such time clocks, and not related to the existence of absolute time.

This text takes the position that neither energy, nor gravity, nor matter, irrespective of how dense an object may be, has any relation to 'absolute time' in the universe. The universe's absolute time clock beats steadily without variability. Absolute time is consistent from one end of the universe to all of the other ends of the universe. Absolute time only moves forward, and unfortunately for all of the science fiction buffs, time does not flow backwards in any set of circumstances other than that contrived by human imagination.

If gravity is actually likened to a phenomenon of fluid dynamics, which is variable throughout the universe, then the phenomenon of absolute time may be the only unyielding perpetual 'constant' in the universe. The means of measurement of true 'absolute time' is likely near impossible. The closest means of measuring absolute time in the universe may present itself as the cyclic behavior of a massive pulsating star. A massive pulsating star would harbor and exhibit enough energy as to not be affected by the fabric of the universe, and would be large enough to be seen, detected and act as beacon by large segments of travelers about the universe. Given that there are numerous pulsating stars, a consensus would have to be determined to designate the prime pulsating star to act as the universal absolute time designating star. Since all stars do exhibit changes in behavior through the course of their existence, a pulsating star could change its behavior. But given the human existence is approximately a hundred years, while a star's existence may be billions of years, changes in the behavior of a star are more than likely not consequential to the need by humans to keep an accurate track of absolute time.

The paradox which comes to mind is that science fiction stories have described the phenomena that a space traveler moving through space would age slower than humans remaining stationary on planet earth. Often it has been described that given a set of identical twins, that for the twin whom ventured off into space for several years, would return to find their identical twin whom had stayed on earth, having aged considerably compared to the aging of the twin whom participated as the space traveler. This alteration in the aging process between the space traveler and the resident on earth, has been floated as a concept tied to the prediction that time will slow down in a clock which is moving at high speed through space, in relation to a clock which is stationary.

In a universe comprised of a luminous aether, the clock which is moving through space is functioning slower than a clock at rest on earth, because the atoms internal to the clock which is in motion is being bombarded at the sub^2 level by quadsitrons comprising the fabric of the universe. A human traveling in space is subject to this same phenomenon. All of the atoms comprising the human body will be bombarded by the quadsitrons flowing at high speed through the atoms and tissues of the body. This constant beating of the protons and neutrons by the accelerated flow of quadsitrons will cause breakdown of atoms and thus molecules comprising various elements of the body. Contrary to what has been previously theorized, it is likely that an unprotected space traveler would age 'faster' than their counterpart residing safe and stationary on earth.

An analogy would be a person standing on earth, in a rain shower, with rain droplets falling from the sky onto their head versus the person's identical twin riding a bike at a fast speed through a similar rainstorm for miles. The person riding the bike would be struck repeatedly across the front of their body by high impact raindrops, where the person standing in the rain would be struck on the head by low impact raindrops. If the velocity of the bike were high enough and the length of the ride were long enough, the person riding the bike, in addition to getting soaked, may actually incur injury such as bruising or muscle aches; while the person standing stationary in the rain shower would be expected to simply get wet if they did not bring an umbrella or raingear with them.

In addition, a space traveler would be expected to be exposed to higher levels of radiation in space, than people residing on the earth. Given the background radiation in space produced by the sun and other stars, human space travelers may age at a much more rapid rate given tissue damage due to radiation exposure. It is imperative, that a means of protecting space travelers be devised before we send humans out on deep space missions for extended periods of absolute time. Later in this text SLIP stream technology is discussed, which is meant to protect humans during faster than light speed interplanetary space travel.

CHAPTER 46

NEWTON'S PARADOX: RE-VISITING THE FIRST LAW OF MOTION

SIR ISAAC NEWTON, LIVED 25 December 1642 to 20 March 1726 and was a noted English physicist and mathematician. He created a number of works, which include Philosophiae Nautralis Principia Mathematica, published 1687. Some Newton's most recognized accomplishments included his Laws of Motion, which were discussed in Philosophiae Nautralis Principia Mathematica.

Isaac Newton had tremendous vision regarding the physics of the universe. For over two hundred years, his principles acted as the underpinning of the world's understanding of physics and math. Often Isaac Newton's descriptions of physics have been referred to as Newtonian Physics.

As translated to English by Andrew Motte in 1850, Newton's three laws of motion include:

Law I: Everybody perseveres in its state of rest, or of uniform motion in a right line, unless it is compelled to change that state by forces impressed thereon.

Law II: The alteration of motion is ever proportional to the motive force impressed; and is made in the direction of the right line in which that force is impressed.

Law III: To every action there is always opposed an equal reaction: or the mutual actions of two bodies upon each other are always equal, and directed to the contrary parts.[138]

Newton's First Law of Motion generally is regarded to mean that an object will remain at rest or in uniform motion in a straight line unless acted upon by some external force. This is a curious statement to make and may have been contrived in Newton's mind prior to his work on his text Opticks: or a Treatise on the Reflections, Refractions, Inflections and Colours. In Opticks, published in 1704, Newton refers to a luminous aether. This luminous aether is meant to penetrate all space and act as the medium for light to transit from one location to another location in the universe. If there exists an aether medium occupying all space, then it is not hard to jump to the assumption that such an aether, no matter how small or how slight, would act as a drag on any object transiting from one location in the universe to another location in the universe. If any form of measurable aether acted on an object, then this concept would interfere with, if not void, Newton's first law of motion. An object traveling through space would eventually slow its velocity related to resistance offered by the existence of the luminous aether.

There are three explanations as to why Newton's First Law of Motion may remain valid in the face of an aether being present in the universe. The first explanation is that Newton felt the particles making up the aether were so small they were inconsequential to the motion of a macro object. Humans generally regard the atmosphere we breath as being inconsequential to our forward locomotion. Only when a wind is gusting against our forward path do we appreciate that the atmosphere may impede our motion.

The second reason Newton may not have referenced the effects of an aether in his First Law of Motion is that the concept of an aether occurred after his First Law of Motion was published and therefore, already engrained in his readership.

The third reason why Newton may have not referenced the effects of an aether is that the luminous aether, theoretically, may behave in a manner different than substances that we are generally aware of as humans. If a submarine were submerged in the ocean, it would be expected that if the submarine's engines were shut off, the submarine initially would continue on its course, but without active engines, the ship would immediately encounter resistance by the water surrounding the submerged vessel and eventually forward motion of the vessel would cease. Once the inertia of the submerged vessel was eroded by the resistance created by the water, the vessel would become subject to the surrounding currents of the ocean waters and move in a direction dictated by the movement of the ocean.

If quadsitrons were simply objects with no charge, it would be logical that the example of the submarine would likely be applicable to any object in space, and that Newton's First Law of Motion would degrade in the face of the presence of an aether. Since an aether is thought to permeate all space, Newton's First Law of Motion would not apply to any practical circumstance since even a vacuum would be thought to be filled with the presence of the aether. But quadsitrons are not without charge. Quadsitrons consist of four charge states to include positive, negative, neutral, and absolute zero. Given the four charge states, individual quadsitrons exhibit a behavior of repelling each other; thus, space comprised of individual quadsitrons does not lock up and become likened to a solid state. 'Space' at the sub-sub atomic level of the quadsitron, always remains fluid, no matter how tightly packed quadsitrons may become due to a force exerted by a highly intense localized gravitational field surrounding each individual quadsitron.

Given that quadsitrons consist of four differing nonreducible charge states, it is possible that as an atom or a structure comprised of atoms moves through space, the atom or structure of atoms meets a resistance in its forward motion through space, but as the atom or structure of atoms passes through a point in space the quadsitrons behind the atom or structure of atoms, due to the inherent charges present in the quadsitrons, exerts a force equal to the resistance to forward motion, which pushes the atom or structure of atoms forward thus conserving the inertia the object possesses. An analogy may be likened to a spring, where a coil absorbs energy and then releases energy. If the charges present in the quadsitrons acted on an object passing through space to cause the object to maintain its inertia indefinitely, then the validity of Newton's First Law of Motion would remain intact.

The only means to prove whether Newton's First Law of Motion was truly valid would be take the test object out of the solar system, well beyond the Kuiper Belt where the effects of the sun and other celestial bodies is neutralized and where there exists no inherent motion quadsitrons and perform motion studies. Maybe someday Mankind will have the technical capability to venture beyond the boundaries of our solar system. Till then, Newton's Three Laws of Motion remain the gold standard of physics.

QUADSITRON MECHANICS: DRIVING FUTURE TECHNOLOGIES

CHAPTER 47

INEXHAUSTIBLE ANTI-GRAVITY ENGINE

IF THE PRINCIPLE OF gravity is indeed governed by the movement of quadsitrons from outer space down through the planet toward the core of the planet, then this action can be exploited to create an inexhaustible engine. If quadsitrons are streaming down from the heavens above headed toward the center of the earth, then diverting the flow of quadsitrons around an object would potentially cause the object to become weightless. A weightless object would rise from the surface of the planet upward. An analogy would be how a scuba diver rises to the surface of a body of water. A diver may increase the quantity of air in his/ her buoyancy vest by diverting air from their air tank to the vest, thus increasing buoyancy, which acts against the density of the surrounding water to cause a rise of the diver in the direction of the surface of the body of water.

An inexhaustible engine would be comprised of a lever that would be capable of rotating upward and downward and a gravity field that would surround the engine. A gravity field intense enough to divert quadsitrons around the lever of the engine could cause the lever to become weightless. See Figure 178.

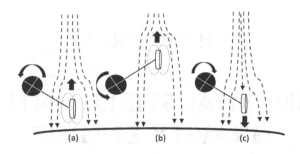

Figure 178

An inexhaustible anti-gravity engine: (a) lever of the engine down prior to gravity field surrounding the lever; (b) gravity field surrounds the lever, making the lever weightless and lever rises; (c) gravity field turned off and the lever drops to the ground due to the force applied by the quadsitrons streaming down from the heavens above.

The lever action of the engine is controlled by turning on and off the gravity field surrounding the lever. With the gravity field turned 'on' the lever rises upward, acting buoyant in relation to the quadsitrons passing through the planet headed toward the core of the planet. With the gravity field turned 'off' the engine's lever is exposed to the flow of quadsitrons streaming from space toward the core of the planet and the lever loses its buoyancy and falls back toward the planet. The result of creating a gravitational field around the lever and then removing the gravitational field, which would cause the lever to rise and fall, creates an engine.

Alternatively, similar to water turbines being used to generate electric power at a dam site, gravity turbines might be used to harness the flow of quadsitrons to power engines. By tapping the continuous flow of quadsitrons moving from outer space toward the core of the earth, gravity turbines could be developed to power engines of all sizes. Gravity turbines would generate no pollution, little heat, and challenge the global use of combustion engines.

CHAPTER 48

INEXHAUSTIBLE
ENERGY ENGINE

A SECOND MEANS OF creating an engine would be to utilize a 'magnet' as an antenna device to harvest useable energy from the universe. A magnet constantly circulates energy surrounding its structure. A magnet is a sink for energy. A magnet will draw in free energy from the surrounding universe until the energy channels positioned around the magnet reach maximum energy capacity for the construct of the particular magnet. Once the magnet's energy channels are at capacity, a magnet exhibits a magnetic field, but has not need to capture additional energy, unless the magnet loses energy. One could bleed off energy from the magnet to create useable energy to drive the motion of an engine.

Electric motor engineers already utilize this theory in principle. As discussed earlier in this text, a magnet passed through a coil of wire will generate a current in the coiled wire. Hypothetically, it is thought that the magnetic field surrounding the magnet exerts a force which pushes the electrons in the 4s orbital of the copper atoms comprising the copper wire. Due to the magnetic field creating a push effect on the atoms of the copper wire, electrons in the 4s orbitals are encouraged to hop from one copper atom to another copper atom, thus creating current flow in the copper wire.

But what if the magnet were simply transferring energy, in the form of quanta, to the copper wire. The magnet, given its unique structural properties, may indeed act as a perpetual energy sink. Magnets may constantly act to absorb any stray free energy quanta which may come within reach of the outer boundaries of the magnetic field surrounding the magnet. Magnetics may automatically absorb and trap free energy until the magnet reaches the capacity dictated by its physical structure. Anytime a magnet loses energy, the magnet simply draws in additional free energy till it reaches its holding capacity. Once a magnet reaches its holding compacity the magnetic is able to exert force to the surrounding environment, but has not need to absorb additional quanta of energy.

Thus, a magnet could be utilized as an energy antenna. Incorporating a gravitational field as a couple to the engine, energy could be transferred from the magnet directly to a motor. As the magnet loses energy, the motor would gain and use the energy. See Figure 179. As the magnet transferred energy to the motor, the magnet would automatically absorb additional energy from the surrounding universe to make up for any loss. In a high energy environment, such as a magnetic energy engine (MEE) exposed to direct sunlight, energy transfer from the magnet to the engine may be continuous, and therefore such a motor may be capable of running nonstop. Even in areas not continuously exposed to the sun, there may still be a substantial amount of free energy which could be harvested from the environment on an uninterrupted basis.

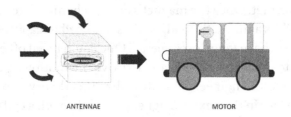

ANTENNAE   MOTOR

Figure 179

Magnetic energy engine, utilizing a magnet to absorb
energy from the environment to power an engine.

An engine utilizing a magnet to absorb free energy from the environment, does not require combustion to create the motion of a motor. In addition, a motor using a magnet as its source of energy will not generate heat to create the power of motion, like gasoline powered engines do by today's design. A motor which utilizes one or more magnets as an antenna to harvest free energy from the environment, will not create elements which pollute the planet we all dearly depend upon for life.

CHAPTER 49

ANTI-GRAVITY FLIGHT

IN ESSENCE, ALL OBJECTS which exist on the Earth are being forced in the direction of the center of the planet due to the flow of quadsitrons streaming from outer space downward through the planet toward the core of the planet. See Figure 180. Creating a gravity field to surround a person, would divert the flow of quadsitrons around the individual, resulting in the individual becoming buoyant, which in the proper circumstances would create lift.

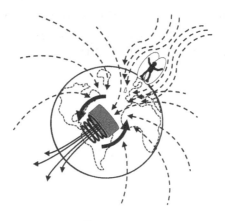

Figure 180

Surrounding a person with a gravitational field creates lift.

If the gravitational field is intense enough, the buoyancy will be great enough to cause the person to rise above the ground and lift off the planet. See Figure 181. Quadsitrons would be diverted around the individual by an intense gravitation field, the diversion of quadsitrons would crowd and eventually fill in the space below the gravitational field, pushing the individual inside the gravitational field upward.

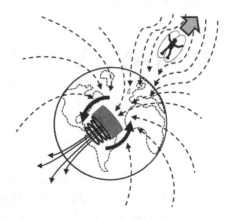

Figure 181

A strong enough gravitational field will create enough lift to cause an individual to separate from the Earth and rise up toward the sky.

Diverting the flow of quadsitrons around an object or a person could create enough lift to launch such entities into space. Imaginative methods to divert and utilize the quadsitron flow streaming toward the center of the planet, could be utilized to increase the speed of the lift phenomenon.

CHAPTER 50

HARNESSING GRAVITATIONAL FIELD TO CREATE THE VEHICLES OF THE FUTURE

IN 1794, ROBERT STREET submitted a patent for the design of an internal combustion engine, the first engine design to use liquid fuel.[139] Though battery operated vehicles are growing in popularity, the vast majority of vehicles in the twenty-first century generally are powered by a gasoline or diesel engine. The combustion of gasoline or diesel fuel inside an engine, produces the power in order to turn an axle, which results in two or more wheels to rotate in relation to the ground, which results in the transport device moving forward or backward in relation to the ground. This utilization of the combustion engine has most certainly accelerated the technological revolution, which has sped forward over the last two hundred years. Arguably, the combustion engine is one of the greatest achievements of modern technology. The worldwide use of combustion engines to power factories and motor vehicles has unfortunately also led to release of enormous amounts of carbon emissions into the atmosphere as a result of burning of fossil fuels. Damage to the ozone layer as well as global warming have been hotly debated, as to the possible cause and effect relationships to the worldwide burning of fossil fuels in combustion engines.

As the earth's fragile environment changes due to whatever cause, and the health of humans becomes threatened due to harmful changes, it may become necessary and inevitable to derive an alternative design to the fossil fuel burning combustion engine.

Diverting the flow of quadsitrons around objects creates lift. If the diversion could be controlled and applied to a vehicle to carry human occupants, then a new, noncombustion form of transportation could be developed. See Figure 182. The weightless car is a concept that could be imagined, conceived, developed, and placed in production.

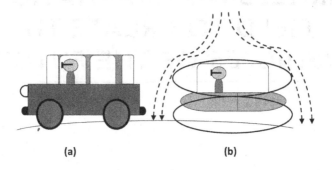

(a)   (b)

Figure 182

(a) Typical land vehicle moves due to wheels rotating overcoming friction created by the surface of the ground resulting in the vehicle moving forward or backward. (b) Creating a gravitational field around a vehicle would create buoyancy, which could be used to move a vehicle without having to overcome friction.

Once the problem of friction is overcome, and a vehicle is weightless, a simple engine, which consumes less fuel than current engine design, could be used to move a weightless vehicle forward. Alternatively, the same engine that creates the divergence of quadsitrons to move around the vehicle, rather than through the vehicle, could be used to divert quadsitrons in a direction opposite of the intended direction of travel, resulting in a form of propulsion, which could move the vehicle in the direction of desired travel.

CHAPTER 51

HARNESSING GRAVITATIONAL FIELD STRENGTH TO LAUNCH SPACECRAFT

GRAVITY IS THE FORCE thought to create a tendency for any and all objects residing on the surface of the earth to physically migrate toward the center of the planet. A rocket is affected by the force of gravity like any other object. The classic means to launch a rocket into space is to build an engine powerful enough to create sufficient thrust to overcome the force of gravity. Successful rocket engines generate lift by rapidly burning tons of fuel, creating a colossal volume of expanding gases below the base of the rocket, which cause the rocket to escape the grip of gravity, lift into the sky and eventually enter outer space. The classic thought is that gravity is some invisible force generated by planet earth which in essence sucks all objects toward the center of the planet.

What if gravity is not some force which pulls from below, but instead is simply the stream of quadsitrons traversing from outer space toward the core of the planet. If gravity is actually a stream of quadsitrons, then the effort of launching a rocket out into space is likened to a boat motoring upstream, against the current. The successful launch of a rocket out into space is simply achieved by creating enough force by exploding volatile gases to cause the rocket to overcome the force of the quadsitrons streaming down from the heavens above. See Figure 183.

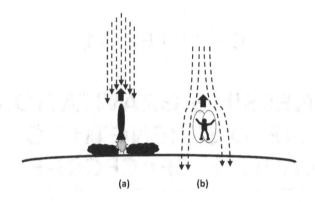

Figure 183

(a) The classic launching of a rocket requires sufficient thrust to counter and overcome the flow of quadsitrons streaming down from the heavens above. **(b)** The use of a gravitational field to create buoyancy offers an alternative means to launch a vehicle into space.

Utilizing a gravity field to divert the quadsitron flow around a vehicle creates lift. Humans could ride inside such a vehicle as passengers and could be launched into space. Instead of utilizing combustion of volatile gases as a means of propulsion, the spacecraft and passengers could simply be made weightless with increased buoyancy creating lift. See Figure 184.

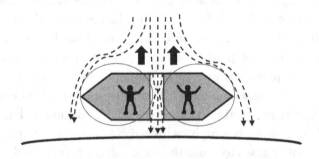

Figure 184

Lift occurs due to a gravitational field diverting the flow of quadsitrons around the spacecraft. In the center of the vehicle a gravity turbine causes quadsitrons to churn, generating upward thrust in the luminous aether.

Gravitational field turbines will propel future spacecraft by channeling the luminous aether to create thrust at the sub-sub atomic level of matter. In the center of the spacecraft could be built a gravitational turbine, which the gravitational force of the turbine would cause quadsitrons to twist and turn on their travel downward toward the center of the earth, which would result in an upward thrust, assisting with the lift capability of the spacecraft. The thrust created by the gravity engine could be regulated to create mobility for the spacecraft.

CHAPTER 52

MAGNETIC BUBBLE TO PROTECT SPACE TRAVELERS

VERY STRONG MAGNETIC WAVES generated by the Earth, wrap around and surround the planet, protecting organic life from lethal radiation that exists in space. The electromagnetic spectrum consists of low frequency waves of energy such as radio waves and mid frequency radiation such as infrared, visible light and ultraviolet light. The electromagnetic spectrum also includes high frequency energy waves such as x-rays and gamma rays. High frequency waves can be damaging to cellular structures of the body by interfering with molecular bonds, and disrupting DNA, and destroying cell structures.

Safe, long distance space travel for organic structures can only be accomplished by exploiting the quadsitrons capability to generate a protective magnetic field to insulate life from lethal radiation encountered during passage through deep space. Irrespective of whether gravity field could be constructed strong enough to lift a space vehicle off the surface of the planet, if there is any hope of deep space travel, we must master magnetism and gravity. See Figure 185.

Figure 185

Artificially creating a protective magnetic bubble around human space travelers is a necessity to protecting life when traveling beyond the shelter of the electromagnetic field that surrounds the Earth.

High energy x-rays and gamma rays stream though human tissues and are able to disrupt chemical bonds. The higher the intensity and the longer the duration of high energy rays, the increased cellular damage that occurs. Very high intensity radiation will kill cells. Lower intensity radiation may disrupt the bonds in the DNA located in a cell's nucleus or mitochondria, resulting in permanent damage to the coding stored in the DNA. Damage to the DNA results in mutations of genes. Such mutations can lead to the cell transcription machinery misreading of the genetic information stored in the DNA. If an extensive amount of DNA is damaged due to exposure to radiation, cell metabolism may be become ineffective. Muscles and body organs may experience premature failure. Cancers and forms of leukemia may occur due to the DNA coding becoming misread by the cell. Following exposure to radiation, death to the body may occur over a short period of time if the radiation dosage is high enough, or at lower doses of radiation death may occur over months to years following exposure.

Organic life cannot survive, cannot flourish outside the protective electromagnetic field surrounding and enveloping the Earth. Unless life is encapsulated in its own protective electromagnetic field, the high levels of radiation emitted by the sun and other stars, which pass through space would end or greatly shorten the lifespan of any organic life traveling through interplanetary space. When Mankind master's magnetism and gravity, then we will be ready to leave the Earth, and successfully explore the surrounding planets and deep space.

CHAPTER 53

CONSTRUCT OF A SUBLIGHT INTRASTELLAR PROPULSION DRIVE

WITHIN TEN LIGHT YEARS distance from our sun exist eleven stars. See Figure 186. The closest stars are a cluster of three stars generally referred to as Alpha Centuri.[140,141] In addition to the Alpha Century trio, there are two binary star systems and four red dwarf stars within ten light years (ly) from earth.[142] Sirius and Luyten 726-8 are the two binary star systems at 8.58 and 8.73 light years distance respectfully. Bernard's star (5.96 ly), Wolf 359 (7.78 ly), Lalande 21185 (8.29 ly), and Ross 154 (9.68 ly) are the four red dwarfs. For the hearty space adventurer, there exist star systems near us to venture forth and explore.

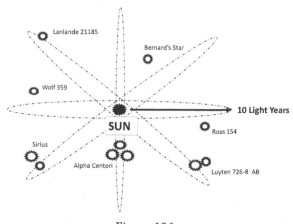

Figure 186

Ten star systems within ten light years distance from our Sun.

The previous discussion of how water and energy are similar in behavior sets the stage for designing an interstellar propulsion drive. Interstellar drive designed for interstellar space travel will take the form of at least two different types of propulsion concepts. One propulsion drive will be likened to a paddle wheel, much like an old Mississippi paddle boat from the 19th-century. The second type of space propulsion drive will take the form of a turbine engine.

The basics of an interstellar space ship can be designed in a variety of ways. The architecture discussed in this text will be one set of plans and the ship will be referred to as the 'Provence'. The interstellar engines will be referred to as 'Doreen Light-speed Interstellar Gravity Hypercoil Turbine' (DORELIGHT) Engines.

The Provence is designed in the manner to take optimal advantage of the function of the magnetic field and the gravitational fields generated by the twin engines. The view from the top of the Provence spacecraft is seen in Figure 187. Top and forward section of the Provence is a command center. Below the command center is living quarters. Behind the command center and living quarters are two DORELIGHT engines.

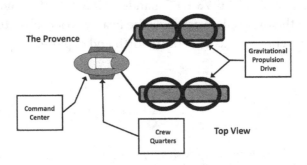

Figure 187

View from the top of the Provence.

From the side view of the Provence one can see the command center sits on top, then extends a shaft down to each engine, while the living quarters sits below and also extends a shaft upward to each engine. See Figure 188.

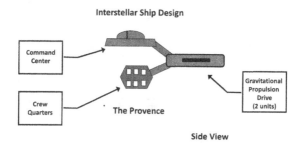

Figure 188

Side view of the Provence.

The front view of the Provence shows the command center on top, the living quarters down below, and the two engines midway between the command center and the living quarters. The two DORELIGHT engines are positions outward to the right side and the left side with angled shafts attaching the command center to the engines from the top and the living quarters from below. See Figure 189.

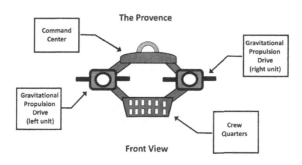

Figure 189

Front view of the Provence.

Sublight speed interstellar propulsion drive would be used for purposes of slower more delicate forward or backward motion of space ship. Sublight speed can be accomplished by creating a gravity field that is strong enough to act as a paddlewheel. The side view of the paddlewheel construction is seen in Figure 190.

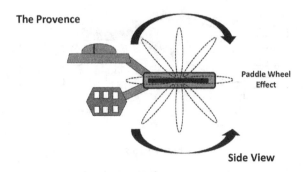

Figure 190

Side view of paddlewheel construct of Provence interstellar space ship.

The gravity force comprising the paddle wheel construct would displace quadsitrons to the rear of the ship creating forward motion causing the ship to traverse in a forward trajectory. See Figure 191.

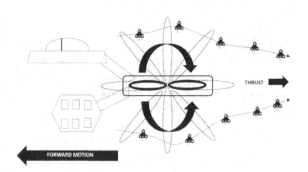

Figure 191

Side view of the Provence showing flow of quadsitrons.

View of the paddle wheel engines as seen above is seen in Figure 192.

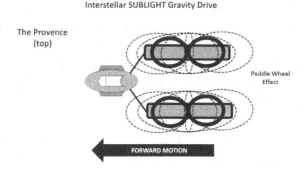

Figure 192

Top view of the Provence showing the rotating
gravitational paddlewheel motion.

The front view of the Provence shows the rotating motion of the gravitation paddlewheel. See Figure 193.

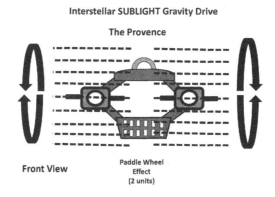

Figure 193

Front view of the Provence showing rotation
gravitational paddlewheel motion.

With the Provence having two engines, having one engine turn faster than the opposing engine will cause the ship to turn to the opposite side. Causing the paddling motion to reverse, would cause the ship to reverse direction. Placing a gravitational baffle rear of the engines could control the upward or downward motion of the ship. By directing the flow of quadsitrons downward with the gravitational baffle would result in upward motion. Directing the flow of quadsitrons upward with the gravitational baffle would result in a downward motion of the Provence.

Any interstellar ship would need to be designed to protect the crew and passengers from the lethal radiation which courses through space. The Provence would be equipped with a magnetic field which would surround the command and living quarters of the ship to protect the ship's crew and passengers. See Figure 194. Gravity fields would be used for propulsion and magnetic fields would be used for crew survival. Local gravity field inside the command center and living quarters would provide a sense of gravity for the crew. Intensifying the protective magnetic field around the ship to the level to match a gravity field may be used as a defense against very high doses of radiation, which may be intermittently encountered in space.

Paddlewheel propulsion system technology would be utilized for slower, more precise movement of an interstellar space ship, such as when the Provence was in the vicinity of a planet or when the ship would dock with a space station.

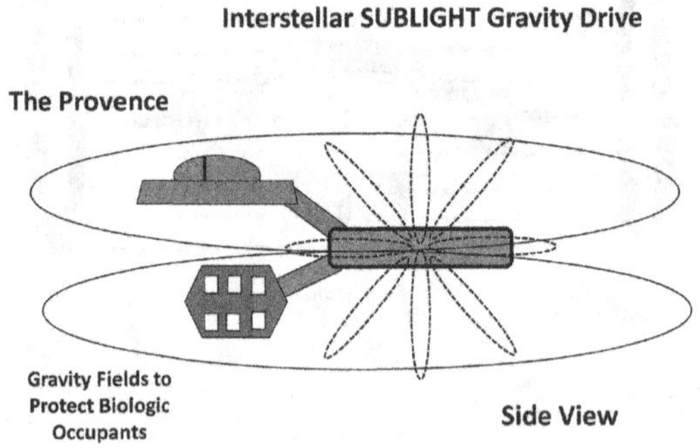

Interstellar SUBLIGHT Gravity Drive

The Provence

Gravity Fields to
Protect Biologic
Occupants

Side View

Figure 194

A protective gravity field would surround the spacecraft.

The protective gravity field would prevent lethal doses of cosmic radiation from damaging the organic structures traveling within the ship.

CHAPTER 54

DORELIGHT ENGINES: DESIGN OF LIGHT SPEED GRAVITY PROPULSION DRIVE

LIGHT SPEED INTERSTELLAR PROPULSION drive is a necessity if humans are to venture beyond the boundaries of our solar system. The distances between stars is so great, humans must achieve at least light speed travel if there is any hope of reaching a star which might provide a habitat that could sustain human life.

Within the radius of ten light years from the Earth's Sun, there exist eleven stars. Three stars are in a cluster referred to as Alpha Centuri; two-star systems are binary systems; four stars are red dwarf suns.

Quadsitrons may behave in a similar manner as air molecules or water molecules. Density of quadsitrons most likely varies throughout the universe. At the same time, the universe itself may be defined by the physical extent to which quadsitrons exist in space. The boundaries of the presence of quadsitrons may coincide with the boundaries of the universe. Like air and water, given different densities of quadsitrons in differing locations about the universe, varying forces of gravity may result in wildly differing flows of quadsitrons.

Capitalizing on the hypothesis that the motion of quadsitrons could be manipulated, turbine like gravity engines could be constructed. Churning forces of gravity utilizing tight coils or hypercoils may result in swirling motion of quadsitrons. Packing quadsitrons through a turbine like device may result in a turbulent flow of quadsitrons, creating thrust, which a spacecraft could use to propel itself from one location to another location in the universe. See Figure 195.

Figure 195

The churning motion of a gravitational field may
result in forcing flow of quadsitrons.

Exertion of the forces of gravity upon quadsitrons resulting in a flow of quadsitrons may be utilized to create thrust for an interstellar spacecraft. As an analogy, jet engines, which power aircraft, force air molecules through turbine engines. The compression of air by the spinning of the turbines within the housing of the engines generates thrust out of the rear of the aircraft. With an aircraft moving forward, the regulation of the flow of air over and under the wings provides lift for the plane.

Utilizing a turbine process, with dense fields of gravity acting as the blades of the turbine, to generate a regulatable flow of quadsitrons to create thrust out of the back of such engines mounted to a spacecraft. See Figure 196. Such a propulsion drive is referred to as Doreen Lightspeed Interstellar Gravity Hypercoil Turbine (DORELIGHT) engines.

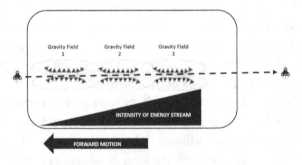

Figure 196

DORELIGHT engines designed to generate a churning
motion within the field of quadsitrons.

In addition to utilizing gravity fields to create turbulence in the quadsitron field to generate thrust to propel a spacecraft, additional gravity fields might be utilized to direct the flow of quadsitrons for steering purposes. Dense gravity fields located at the output end of the DORELIGHT engines, could be constructed like the rudder of a boat to direct the thrust created by the engines to assist the crew of the spacecraft in maneuvering the course of the ship. See Figure 197.

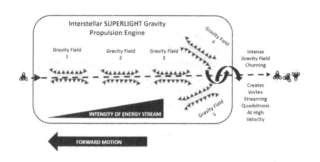

Figure 197

Doreen Light-speed Interstellar Gravity Hypercoil Turbine engines will have coiling gravitational fields set up to direct flow of quadsitrons out the back of the engines.

The Provence would be equipped with both paddlewheel type engines and DORELIGHT engines. Picture of the Provence is provided in Figure 198. Paddlewheel engines would be utilized for slower maneuverability such as in the vicinity of a planet, while DORELIGHT engines would be used to create the faster flight speed required for travel between stars.

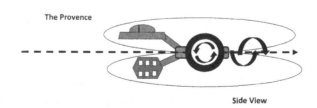

Figure 198

The Provence would be equipped with two DORELIGHT engines.

Therefore, it is understood that for most elements quadsitrons pass through the element or molecular object. See Figure 199. Magnetic substances channel energy. Quadruthenium, a theoretical substance, may funnel and churn quadsitrons.

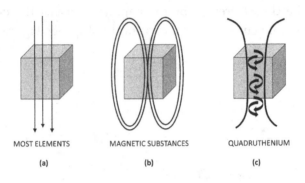

MOST ELEMENTS   MAGNETIC SUBSTANCES   QUADRUTHENIUM

(a)   (b)   (c)

Figure 199

Illustration depicts (a) quadsitrons passing through most substances, (b) magnetic substances channeling quadsitrons and (c) quadruthenium, a hypothetical substance capable of churning quadsitrons.

Obviously, quadruthenium is a theoretical, possibly a theatrical substance, to give life to the concept that gravity engines can be 'built' and constructed with such intense energy as to create a propulsion drive to propel starships through space, much like turbine engines propel airplanes through the sky. The technology required to harness energy dense enough to generate a gravity effect strong enough to exert force of the movement of quadsitrons is speculative at best at this time.

Magnetic materials are substances which exert a magnetic field about their boundaries. Magnetic fields realign the position of quadsitrons in space about one or more axes of the quadsitrons, which results in the trapping and channeling of the flow of energy. Magnetic fields are not strong enough to appreciably alter the movement of quadsitrons in space from one location to another location.

Bar magnets and the magnetic field they exhibit are a known entity. How a magnetic substance is able to realign the special orientation of quadsitrons to trap energy is a subject which requires further study and is beyond the scope of this text. A substance containing the molecular construct to create a churning motion of quadsitrons, is arbitrarily referred to as quadruthenium in this text.

The even mention of a hypothetical substance such as quadruthenium assures the reader that this text ventures into the realm of science fiction. The past few decades of human technologic advancement has shown that each generation's standard of technology often starts as a form of fiction from a preceding generation of dreamers. Once the construct of a magnetic field has been defined, it may be possible to define how molecular structures exist in nature or how molecular structures can be built in order to manipulate gravitational fields to facilitate the engine technology to send spacecraft to distant stars.

CHAPTER 55

FASTER THAN LIGHTSPEED: SUPER-LIGHT INTERSTELLAR PROPULSION (SLIP) STREAM TECHNOLOGY

SUPER-LIGHT INTERSTELLAR PROPULSION (SLIP) Stream Technology is a necessity for traversing the vastly long distances of interstellar space travel. Such technology, though deemed for faster than light travel, may in fact be a necessity in the near future to allow humans to safely travel between planets, such as from Earth to Mars. In addition to protecting space travelers from lethal doses of radiation, SLIP Stream Technology would be utilized to prevent space travelers from aging faster than their human counterparts left behind on Earth. Albert Einstein had predicted humans participating in space travel would age less than counterparts remaining on Earth, but in reality, space travelers without protection from the luminous aether, would more than likely age faster.

Since the dawn of mankind pondering interstellar space travel, there has existed the controversy of whether an object could travel faster than the speed of light. The thought was that as an object approached the speed of light, the object would collapse and transform into pure energy. Additionally, it was thought that at a velocity equal to the speed of light, time would grind to a halt. Also, for the last 132 years, it has been professed that light traveled from one location in the universe to another location in the universe without the aid of a transfer medium.

For a significant portion of human history, it was believed that no object could travel faster than the speed of sound. Then on October 14,

1947, Chuck Yeager flying the X-1 aircraft, broke the sound barrier.[143] Now we refer to jet or rocket speed as multiples of the speed of sound when it is referred to a Mach speed. Since 1947, it has been common knowledge that the speed of sound is simply a physical phenomenon and not a barrier that limits how fast humans can travel.

Science fiction has proposed that it is possible for a spacecraft carrying humans to travel faster than the speed of light. Warp speed has been considered by science fiction authors to represent travel faster than the speed of light.[144] Like the speed of sound, the haunting question which tugs at our imagination, is the speed of light simply a physical characteristic of electromagnetic energy and therefore a barrier which can be surpassed, or not.

If light utilizes quadsitrons as a luminous aether facilitating transfer of energy in the form of a wave from one location in the universe to another location in the universe, then this is simply a physical characteristic of electromagnetic energy and not necessarily a physical barrier.

Utilizing the knowledge of a luminous ether will lead to developing and launching spacecraft from the surface of the earth out into space without combustion, but instead by utilizing a magnetic field to create a bubble to counter gravity and generate weightlessness (lift) by deflecting the energy being channeled downward onto the earth's surface, which will result in buoyancy within the earth's gravitational field. Gravity field turbines will propel future spacecraft by channeling the luminous aether to create thrust at the sub-sub atomic level of matter. Safe, long distance space travel for organic structures can only be accomplished by mastering the channeling of quadsitron to generate a protective magnetic field to insulate life from lethal radiation encountered during passage through deep space. Faster than the speed of light travel would be possible by utilizing gravity fields to create a galactic channel stream, whereby the gravity field would push the quadsitrons comprising the luminous aether out of the forward path of a space ship thus eliminating the solar wind effect that would normally act as a resistance at the surface of the protons and neutrons comprising the body and contents of such a spacecraft.

The view of the Provence is seen from above in Figure 200 with a pair of Doreen light-speed interstellar hypercoil gravity turbine engines. The illustration suggests a tornado effect per each engine due to the hypercoiling gravity turbine propulsion drives.

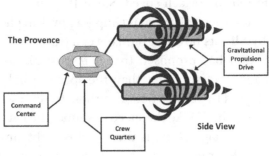

Figure 200

Two DORELIGHT engines illustrated with a tornado effect creating thrust.

The flow of quadsitrons can be seen passing through the two Doreen light-speed interstellar gravity hypercoil turbine engines creating thrust, but also quadsitrons flow through the crew cabins. This is where a revision must occur. This text previously took the position that the flow of quadsitrons through the boundaries of an atom did not result in any consequence to the movement of a particular atom. This subject was touched upon when the Michelson and Morley experiment of 1887 was discussed. The examination of Michelson and Morley's experiment brought to light the distance between the nucleus of an atom and the electron orbits was so expansive in comparison to the size of a single quadsitron, that the vast majority of quadsitrons flowing through an atom at any given time would pass through the boundaries of the atom with their transit trajectory undisturbed. Further, due to the fact that most quadsitrons would pass through an atom without an appreciable change in trajectory, no wake would be formed, except at the surface of the protons and neutrons comprising the nucleus as the atom. Expanding this notion to molecular structures, where a molecule is comprised of two or more atoms, again the majority of quadsitrons occupying space, transiting through molecular structures, pass through such structures unimpeded. A very small ratio of quadsitrons interact with the nucleus of each atom of a molecule.

Now the edge of imagination is approached. To further study the phenomenon of quadsitrons passing through atoms unimpeded in their trajectory, one considers the concept on a planet wide scale. A planet is comprised of an inestimable trillion upon trillions upon trillions of molecules, with each molecule comprised of two or more atoms. The nucleus of each atom may be comprised of one proton (hydrogen), or up to 118 protons and 176 neutrons (Oganesson).

Michelson and Morley, took the position, that there should be a substantial wake generated in the luminous aether by planet Earth as this massive planetary molecular structure orbits around the sun at a velocity of 67,000 miles an hour. Intuitively, this would seem to be a very logical position to take; for all practical purposes, Michelson and Morley's hypothesis that a wake occurs due to the presence of a luminous aether is correct; but at a substantially lower level, in the realm of subatomic space rather than macro space. The failing in the proving the hypothesis lies in the means, by which, the attempt to prove the existence of the wake in the aether was carried out. With light being comprised of individual packets of energy, referred to as quanta (quantum mechanics), a beam of light is comprised of numerous packets of energy. The majority of the quanta of light are going to utilize the quadsitron transfer medium, which the quadsitrons themselves flow through atoms comprising the atmosphere generally unimpeded, and thus the majority of the quanta of light will pass through the air molecules generally unimpeded. In addition, the earth has been orbiting the sun at steady state for presumably millions of years, and any resistance created by a wake would have been diminished to the lowest effect over time. Thus, if one is attempting to study the wake created in the luminous aether by planet earth orbiting the sun, an alternative means of experiment would need to be pursued.

To revisit Michelson and Morley's experiment of 1887, the two scientists theorized that if two beams of light were used, each beam of light pointed at a different target, the axis of the travel of light 90 degrees with respect to each other, that if a wake in the luminous aether existed, then the beam of light affected by the wake in the luminous aether would reach its target slower than the second beam of light not affected by the wake; this was found not to be true, for both beams of light reached their targets at the same time. Again, taking into account quantum mechanics, where light is comprised of quanta, some of the quanta of both beams of light would reach their respective targets at the same time, thus making the results of the experiment appear to favor the opinion that a luminous aether does not exist.

But if the perspective is taken, that the only wake created in the luminous aether as planet earth orbits the sun is detectable only on the surface of the protons and neutrons of each atom, then Michelson and Morley's experiment could be re-visited by conducting the identical experiment, but instead of measuring differences in speed of travel to each of the two targets, the quantity of light each target absorbs over a given period of time is measured. The differences in amount of light absorbed by each target over a given period of time may be infinitesimally small, but if a wake exists, should, at some level, be measurable. Alternatively, one could design an experiment where the air through which the laser beams pass is denser or the air is occupied by larger molecules or both. The quanta of light passing directly by the nucleus of the atoms comprising the atmosphere should have their trajectories disrupted by the wake phenomenon occurring on the surfaces of the protons and neutrons of the atoms in the atmosphere, thus some of these quanta of energy should be lost and thus will not be accounted for by the target affected by the wake in the luminous aether. Thus, the existence of the luminous aether could be supported by an optics experiment that utilized two beams of light directed at two differing targets which could absorb the light, the axis of travel of both beams of light ninety degrees with respect to each other. The beam of light affected by the luminous aether would show a decline in absorption of quanta of energy in comparison to the beam of light not affected by the wake in the luminous aether created by the earth orbiting the sun.

The reason for clarifying the existence of the wake in the luminous aether, at the subatomic level, is to expand the frontier of space travel into super light speed. The barrier posed by the speed of light is simply light depends upon the transfer medium of quadsitrons to travel from point A to point B in the universe. Similar to sound passing through air or sound passing through water (slower) or sound passing through a metal bar (even slower), the speed of travel of the energy is dependent upon the substance the energy is being transmitted through.

Most likely the 'light speed' is an arbitrary barrier. Similar to the arbitrary 'sound barrier', it is likely objects may be able to venture faster than the speed of light. Faster than light travel may simply require imaginative engineering.

It has been proposed that upon achieving the speed of light, mass is transformed into pure energy and time is said to stand still. The notion that mass is transformed into pure energy at the speed of light seems to facilitate understanding of Albert Einstein's equation $E = MC^2$, where mass and energy appear to coincide and seem to be transferable in the study of the atomic world; yet this famous equation does represent mass (M) and the speed of light (C) as two separate entities. Proof of the concept that mass transforms into pure energy at the speed of light will be determined once mankind is able to construct spacecraft which achieve the speed of light. Currently, the fastest spacecraft humanity has constructed is an outbound satellite traveling at 36,000 miles an hour; the speed of light is approximately 720 million miles an hour.

The scope of this text will treat the speed of light as simply an arbitrary speed boundary. For the moment, it will be assumed that achieving the speed of light does not automatically turn mass into pure energy, and it will be assumed that time does not stand still; for either occurrence would be very disruptive for human space travelers. If space travel at a velocity that is faster than the speed of light was achievable, how could this be accomplished.

Part of the answer to faster than the speed of light travel puzzle diverts back to the wake affect created in the luminous aether. Again, it is postulated that the majority of quadsitrons flow through atoms unimpeded, and it is further postulated that a very small number of quadsitron are deflected by the protons and neutrons comprising the nucleus of atoms creating a wake limited to being expressed on the surface of the nucleus of atoms.

Thus, one means that faster than the speed of light travel could be achieved would be by eliminating the drag force created by quadsitrons being deflected by the protons and neutrons present in the nucleus of atoms comprising molecular structures. With zero possible drag to slow down a spacecraft, that is no measurable wake, such a spacecraft could travel as fast as the gravity engines could be constructed to propel the ship.

The Provence as designed up to this point would demonstrate flow of quadsitrons through a pair of DORELIGHT engines as well as through the spacecraft itself. See Figure 201.

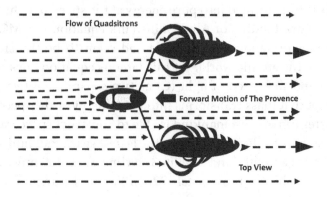

Figure 201

Flow of quadsitrons through the pair of DORELIGHT
engines and the crew's quarters.

A magnetic field would be generated to protect the components of the spacecraft and the human passengers from the lethal effects of radiation encountered in space travel. See Figure 202. The magnetic field protecting the occupants of the spacecraft may be referenced as a magnetic bubble.

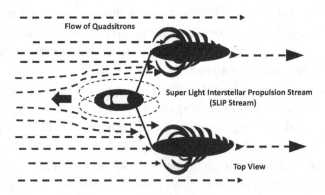

Figure 202

A magnetic bubble surrounds the living and working
quarters of the Provence spacecraft.

A snow train plows railway train tracks in the winter months to keep trains running in snowy regions of the world. See Figure 203. The shovel on the front of the snow train is designed to divert the snow off the train tracks.

Figure 203

Snow train plowing snow off train tracks.

The Provence could be equipped with a forward reaching gravitational field. See Figure 204. The forward reaching gravitational field could be made strong enough to force quadsitrons out of the path of the ship.

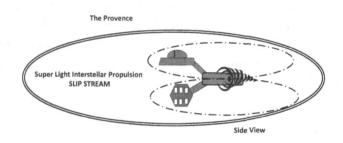

Figure 204

Forward reaching gravitational field.

To achieve faster than the speed of light travel, the Provence interstellar spacecraft would be equipped with a forward reaching gravitational field utilized to divert quadsitrons out of the path of the living and working quarters of the spacecraft, but allow flow of quadsitrons into the pair of DORELIGHT engines. See Figure 205.

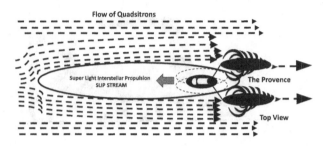

Figure 205

The Provence equipped with forward reaching gravity field to divert quadsitron flow out of the pathway of the spacecraft.

Diverting quadsitrons out of the path of such an interstellar spacecraft as the Provence, is referred to as Super Light Interstellar Propulsion (SLIP) Stream technology. How fast the Provence could travel between stars is only dependent upon how technologically advanced the DORELIGHT engine technology can be developed. Therefore, theoretically, faster-than-light speed should be possible. The SLIP stream protects any drag from being placed on the protons and neutrons comprising the atoms which make up the ship or the crew as they pass through the quadsitrons of the luminous aether.

ALTERNATIVE OPTION: THE OCTATRON

Substitute Elemental Particle Designs

THE ASTUTE READER, PONDERING the key concepts of this book in the three-dimensional viewing space between the posterior parietal lobe and the anterior portion of the occipital lobe of each side of their brain, would have by now started to question the validity of the construct of the quadsitron. After initially thinking that such a particle just could not possibly exist, the reader would then come to the conclusion that a particle with four poles actually makes sense. A four-pole particle, each pole 120-degrees from other poles, each pole being either positive, negative, neutral or absolute zero. Such a four-pole particle comfortably assists in modeling the sub-sub atomic universe and provides a root explanation for the existence of natural cosmic phenomenon seen around the universe.

Obviously, a four-poled quadsitron is not the only possible design for the nonreducible particle comprising the fabric of the universe. There exists the possibility that the universe's most elemental particle is comprised of more than four poles. Possibly the universe's most elemental particle is comprised of six-poles, eight-poles, twelve-poles, sixteen-poles or even more poles. Eight poles or doubling the number of poles described with a quadsitron seems the most likely alternative design to the four-pole concept. See Figure 206. An eight-poled elemental particle might be called an octatron. The eight poles of an octatron may demonstrate 60-degrees of separation, rather than the 120-degrees of separation seen with a quadsitron.

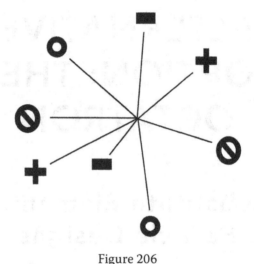

Figure 206

The eight poles of a proposed octatron.

An octatron would likely consist of two positive poles, two negative poles, two neutral poles and two poles which represented absolute zero. Since the perpetual dynamic flow of the universe is dependent upon the elemental particle having four distinctly different charged poles, an eight poled elemental particle has merit. The more poles the elemental particle contains, seemingly the less chance the elemental particles would bunch together, become stuck together and freeze up, resulting in a portion of the universe to in essence crystalize.

Though the possibility of a quadsitron being comprised of some multiple of four greater than 1 x 4 poles, such a discussion is beyond the scope of this book and would be even more difficult to prove than the design of the four-poled quadsitron presented in this book. The matter of the difference between a proton and neutron as described earlier in this text is that of a single surface layer of quadsitrons covering the proton to account for the mass difference. With the quadsitrons having a construct of four poles with the 120-degree separation of the four poles, facilitating a quadsitron being capable of striding another quadsitron to make observation of Beta decay convincingly feasible in a practical sense, presents the most plausible version of an elemental particle design. Still, the possibility that the primary particle has more than four-poles exists, and at some point in the future, may need to be considered in further detail.

SECTION IX

UNIFIED PRINCIPLES OF GLEAM²

SUMMARY

GLEAM²'S PRIMARY UNIFYING CONCEPT

THE PRIMARY UNIFYING CONCEPT discussed in this text is that the universe is comprised of two, and only two elements. The universe is comprised of the two nonreducible elements of (1) ENERGY and (2) QUADSITRONS. At the sub-sub atomic level, the universe is a sea of quadsitrons, through which energy ebbs and flows, and under certain circumstances energy becomes detained. Protons, neutrons, and electrons are constructed of quadsitrons, which have either trapped or are under the influence of energy. Atoms are constructed of the interplay between protons, neutrons, and electrons. Atoms are distinguished in the Periodic Table by the number of protons and neutrons compacted into the center and the number of electrons orbiting the atom, per the conventional Electron Shell Model of the atom; the Lamniscate Model of the Atom provides a differing atomic design scheme for the atom. Molecules are comprised of constructs of atoms. In essence, all that we see and experience with human senses is constructed from the fabric of the universe, crafted with combinations of quadsitrons and energy.

Beta decay of a free neutron has been observed to result a proton, an electron, a neutrino, and a measure of energy. Here, the neutrino is theorized to be composed of two nonreducible components, that of energy and that of two quadsitrons. It is observed that each quadsitron is an elemental sub-sub atomic particle comprised of four poles, opposed

120 degrees to each other, representing four differing charge states to include positive, negative, neutral, and absolute zero. Demonstrated by mathematical derivation it is shown in this text the mass (weight) of a single quadsitron is 9.5075301706 * 10^{-38} kg, and the diameter is calculated per both volume of a neutron and volume of a proton means to be 8.557941465 * 10^{-19} m. The calculated mass and diameter of a quadsitron is applied to the structure of a proton and neutron to show the loss of a single layer of quadsitrons results in a neutron transforming to a proton.

Prior to 1887, the popular scientific belief was that a luminous aether filled the universe. Sir Isaac Newton promoted the concept of the luminous aether, where he referred to light utilizing a medium for transfer. Light was thought to utilize the luminous aether to travel any distance from one location in the universe to another location.

In 1887, Albert A. Michelson and Edward W. Morley's efforts discounted the idea that a luminous ether filled the space between heavenly bodies. Michelson and Morley theorized that if a luminous ether existed, then the earth would generate a solar wind as the mass of the planet passed through such an ether in space while the planet orbited the sun. Michelson and Morley theorized that if such a solar wind existed on the surface of the planet, then this solar wind would disrupt the luminous ether resulting in a measurable effect on the speed of light on the surface of the planet. Michelson and Morley conducted an experiment in which they measured the speed of light of two beams of light perpendicular to each other, and found there was no difference in the speed of light of the two beams of light. Michelson and Morley concluded that since there was no difference in the speed of light between the two beams of light perpendicular to each other, no luminous ether existed.

The radius of an electron per CODATA 2014 is 2.8179403227 * 10^{-15} m. The diameter of an electron is therefore 5.6358806454 * 10^{-15} m. The value of Bohr's radius per CODATA 2014 is 5.2917721067 * 10^{-11} m. Therefore, the number of electrons that can fit side by side along the radius of a hydrogen atom is calculated to be 18,778. The calculated diameter of a quadsitron is 8.557941465 * 10^{-19} m. The number of quadsitrons that could fit side by side along the radius of a hydrogen atom is calculated to be 61,834,637.

The diameter of the earth is 1.2742 *10^7 m. The distance from the sun to the earth is 1.496 * 10^{11} m. The number of earth-like planets that could fit in the distance from the sun to the earth can be calculated to be 11,740.

More electrons would fit between the proton and the electron in a hydrogen atom than the number of earths that would fit between the earth and the sun in our solar system. From the perspective of the proton, there exists considerable empty space between the proton and the electron in a hydrogen atom. Rutherford described the atom as being comprised of mostly empty space.

The mean orbital velocity of the earth is reported as 29.78 km/s. All the mass comprising the earth is traveling in orbit around the sun at a mean velocity of 107,208 km/hr. By comparison, the fastest manned aircraft is documented as being the North American X-15 with a max speed of Mach 6.70, which is approximately 7,200 km/hr. Given Einstein's Theory of Relativity, each proton on the planet, with a mass equal to 1.7592599513 * 10^{10} quadsitrons is traveling at a speed of at least 29.78 km/s through the luminous aether, this aether comprised mostly of loose quadsitrons; other forces acting on the earth related to the earth's position in the galaxy and universe in general may also be affecting the true velocity and momentum of the planet in relation to the luminous aether.

If quadsitrons, with a diameter of 8.557941465 * 10^{-19} m, comprise the luminous aether of the universe, then as each individual atom comprising the earth passes through space, only a very small portion of the total volume of the atom makes any contact with the luminous ether. The vast majority of the quadsitrons comprising the luminous ether pass unimpeded through the space occupying volume of each atom comprising the earth. Contrary to the assumption made by Michelson and Morley in 1887, and given Einstein's Theory of Relativity, there would be no measurable solar wind present on the earth's surface as the planet traverses its orbits around the sun. To measure the presence of a solar wind generated by the earth passing through the luminous ether, one would need to measure the solar wind on the surface of a proton (or neutron) as the proton (or neutron) passes though the luminous aether. Michelson and Morley's experiment's results were correct for the level of understanding of the atomic structure of the day; but given a more advanced science of our day, the experiment's results do not discount the presence of a luminous ether. As Sir Isaac Newton promoted, a luminous aether likely exists constituting the underlying fabric of the universe. As noted in Newton's text *Opticks* 'And so if any one should suppose that Aether (like our Air) may contain Particles which endeavor to recede from one another (I do not know what this Aether is) and that its Particles are exceedingly smaller than those of Air....'.

Reinstating the theory of the luminous aether is critical in explaining the existence of the behavior of light as a wave, as well as the behavior of other wave energies comprising the electromagnetic spectrum. Free energy exists such as quanta moving as waves with energy levels depending upon the frequency and wavelength as represented by the electromagnetic spectrum. Energy that flows at low frequencies, such as sound, is thought to vibrate molecules that act as the transfer medium. Energy that flows at high frequencies is currently thought to have no medium of transfer. Quadsitrons may act as the means of all energy transfer. Vibration of molecules as energy transfers from one location to another may be a secondary phenomenon. Both energy and quadsitrons being nonreducible would explain why light is capable of traveling vast distances through space.

The presence of a luminous aether explains the phenomena of a black hole, where a galaxy is a hurricane-like cloud of quadsitrons and light follows the behavior of its transport medium as quadsitrons enter, then pass through the twisting funnel vortex created at the center of a galaxy. The intense churning forces within the galaxy's vortex disassembles matter as well as creates matter. The brilliance of a quasar is light being seen exiting the vortex of a galaxy. The luminous aether explains how, in the proximity of a bar magnet, energy is trapped in channels created by aligned quadsitrons to create the behavior of a magnetic field surrounding a bar magnet. Similarly, the luminous aether is the medium that is created by aligning quadsitrons into channels of energy that surround the earth to make possible the electromagnetic field which protects life from the lethal radiation emitted by the sun. A luminous ether explains gravity, by providing a medium that channels energy to surround the planet forcing objects down onto the planet's surface by the flow of quadsitrons pushing down on objects from above to create the phenomenon of weight; rather than the traditional theory of an as of yet unexplained attractive force generated from within the planet.

In his address to University of Leiden on 5 May 1920, regarding the luminous ether, Einstein closed his remarks with, 'But this ether may not be thought of as endowed with the quality characteristic of a ponderable media, as consisting of parts which may be tracked through time. The idea of motion may not be applied.' Paradoxically, the existence of a luminous aether explains the phenomena of Albert Einstein's Nobel Prize winning Photoelectric Effect, whereby light passing by a body in space with a large gravitational field would be attracted toward such a body. The presence of a luminous aether, acting as a transport medium for light, would bend light toward a heavenly body that exhibited a higher density of quadsitrons surrounding such a body due to light following a more intact or denser path of its transport medium.

Utilizing the knowledge of a luminous aether will lead to developing and launching spacecraft from the surface of the earth out into space without combustion, but instead by manipulating the gravity field to create a bubble to counter gravity and generate weightlessness (lift) by deflecting the energy being channeled downward onto the earth's surface, which will result in buoyancy within the earth's gravitational field. Gravity field turbines will propel future spacecraft by channeling the luminous aether to create thrust at the sub-sub atomic level of matter. Safe, long distance space travel for organic structures can only be accomplished by mastering the quadsitrons and energy to generate a protective magnetic field to insulate life from lethal radiation encountered during passage through deep space. Faster than the speed of light travel would be possible by adding sufficient quanta of energy to the magnetic fields to create a galactic channel stream, whereby the now resultant gravity field would push the quadsitrons comprising the luminous aether out of the forward path of a space ship thus eliminating the solar sub^2 atomic wind effect that would normally act as a resistance at the surface of the protons and neutrons comprising the body and contents of such a spacecraft. Generating such a galactic channel stream or SLIP stream would facilitate faster than the speed of light travel.

In conclusion, the universe could be considered to be comprised of and defined by two fundamental, uniform, nonreducible elements, that of energy and that of a luminous aether comprised of quadsitrons. This work supports the presence of a luminous aether constituting the underlying fabric of the universe. The calculations presented demonstrate the luminous aether to be comprised of quadsitrons, which in the

317

presence of trapped energy form to create all matter. Quadsitrons are defined as being uniform elemental sub-sub atomic bodies configured with four poles that of positive, negative, neutral, and absolute zero, with a calculated mass of $9.5075301706 * 10^{-38}$ kg and a calculated diameter of $8.557941465 * 10^{-19}$ m. A quanta of energy is $1.7 * 10^{-20}$ joules. From these two entities, quadsitrons and quanta of energy, all else in the universe is created.

CLASSIC PHYSICS VERSUS QUADSITRON PHYSICS

TO ANSWER THE QUESTIONS posed in the first chapter of this text utilizing quadsitron-energy physics, the following table is provided:

Physics Dilemma	Conventional Physics	Quadsitron-Energy Physics
1.Gravity	Unexplained generic force draws objects down toward the core from inside the planet.	Energy dense enough circulating through the planet to the core, resulting in a flow quadsitrons toward the center of a planet.
2.Speed of light	Force of gravity can bend light, slow it down, and even stop light, trapping light in a black hole.	The speed of light is a constant relative to the motion of quadsitrons comprising the fabric of the universe. The speed of light is simply a measure of how fast energy moves through the quadsitron aether in the form of a sine wave.

3.Mass and speed of light	At the speed of light, an object has no mass.	An electron, which travels at the speed of light, consists of 463 rings of quadsitrons, which have mass, surrounding a quanta of energy.
4.Behavior of the electromagnetic spectrum (EMS)	No explanation why light occurs in a wave pattern, when light crosses space in a vacuum, where for all practical purposes no form of matter exists.	Aether comprised of quadsitrons acts as transfer medium facilitating the transfer of energy in the form of waves, from one point in the universe to another point in the universe.
5.Light as a wave, light as a ball	No explanation.	Aether comprised of quadsitrons acts as transfer medium facilitating wave forms for the entire EMS; solid nature of light is related at times to the mass of the quadsitron which encircle quanta of energy or the density of the transfer medium light utilizes for transfer.
6.Light from a distant star reaches earth	Light transfers across vast distances through a vacuum.	Light traverses distance in space by utilizing the luminous aether, which is comprised of irreducible quadsitrons. Since energy and the quadsitrons are both irreducible, the light energy must continue to move forward across space until absorbed.

7.Light reflected off an object as color	Light bounces off the surface of atoms to be seen by someone viewing the object.	The radiant energy of light strikes an atom, as the atom absorbs the energy, the atom vibrates, this vibration ripples out and away from the atom by means of the luminous aether, striking the eye of the viewer, with the retina of the viewer interpreting the resultant radiation as color.
8.Electron orbitals	Complex orbital shapes, with no explanation as to how the orbitals are formed or why electrons do not crash into each other.	Mass of protons and electrons are organized and connected with figure 8 orbitals.
9.What is magnetism?	A magnetic field, no explanation as to how this forms around a bar magnet or a wire.	Channels of quadsitrons surrounding a magnetic substance, which trap free energy in fields about the substance.
10.Electromagnetic field which protects the earth	No clear explanation.	Earth acts as a huge bar magnet, which creates channels of energy flow, which extend out into space, which diverts energy emitted by the sun away from the surface of the planet. Note: Magnetic energy is denser than EMS energy because magnetic energy surrounding the earth deflects the EMS radiation from the Sun.

11.Beta Decay of Neutron to a Proton	Energy releasing event which happens.	The surface layer of quadsitrons peels off the Neutron, emitting energy, and releasing enough quadsitrons to form an electron, resulting in the body of a proton, an electron and a neutrino.
12.Construct of a proton	Comprised of oddly formed quarks.	Comprised of a uniform arrangement of quadsitrons bound by energy.
13.Construct of a neutron	Comprised of oddly formed quarks.	Comprised of a proton covered with a single layer of quadsitrons bound by a weak flow of energy across the surface of the neutron.
14.Construct of an electron	Energy traveling at the speed of light.	Rings of quadsitrons, which surround a quanta of energy.
15.How is the radii of gold smaller than the radii of sodium?	Given the Electron Shell Model of the Atom, the higher density of the nucleus of gold is suspected to draw the complex orbitals closer to the center of the atom. It is unclear as to how electrons would not crash into each other or the atom's nucleus.	Lamniscate design of the atom demonstrates a series of figure 8 orbitals positioned around the center of the atom, where electrons do not interfere or collide with each other or the nucleus of the atom.

16.Why a black hole does not explode?	Gravity so intense it traps light. Uncertain why eventually this does not explode since such a concept would continuously build up energy without a method of releasing energy.	The vortex of a galaxy, which acts like a galactic hurricane, which funnels quadsitrons through, which results in light following the flow of its transfer medium, with light flowing into one end, the black hole, and being emitted on the opposite end, appearing as a quasar. Energy is constantly being sucked into the vortex and subsequently released, there is not a build up of energy, which would cause instability.
17.Light draws toward gravitational body	Due to gravity, an unexplained generic force.	Density of quadsitrons higher closer to a gravitational body, which increases the density of the transfer medium, which light tends to follow the optimal density of its transfer medium.

APPENDIX

CALCULATING THE MASS AND DIAMETER OF A QUADSITRON

THE ILLUSTRATION OF A quadsitron is seen in Figure 207.

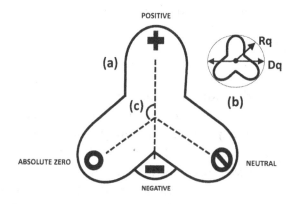

Figure 207

Illustration of a quadsitron.

Calculating the mass and diameter of a quadsitrons is as follows:
Defining the radius of a proton and a neutron.

The objective of this study is to determine the mass of a quadsitron represented as 'Mq' and the diameter of a quadsitron represented as 'Dq'. It is reasoned that neutrons and protons are approximately spherical in shape.

The exact measurement of the radius of a proton has been so elusive that the term 'proton radius puzzle' has been applied to this phenomenon.[145,146] The size of a proton has been measured by at least two differing approaches, which include electron-proton scattering and muonic hydrogen Lamb shift experiments. Electron-proton scattering experiments where a proton is bombarded by electrons estimates the proton root mean square charge radius to be 0.8751(61) fm which is listed in 2014 CODATA.[38] The muonic hydrogen Lamb shift experiments, where proton is engaged by a muonon, 200 times the weight of an electron, estimates the proton radius to be 0.84184(64) fm.[146] The radius of the proton per 2S-2P transmission frequency of muonic hydrogen is 0.84087(39) fm.[147] The Baryon Summary Table in the Review of Particle Physics lists the charge radius of a proton as 0.877 fm.[148] Sha Yin Yue has calculated the proton radius to be 1.112772961016 * 10^{-15} m and the neutron radius to be 1.113284057367 *10^{-15} m.[149] Yue utilizes the following equations:

Rn = radius of a neutron = k * Qp * Qe/ ((Mn-Mp) * C^2)   (1)

Rp = radius of proton = $(Mp/Mn)^{1/3}$ * Rn   (2)

Re = $(Me/Mn)^{1/3}$ * Rn   (3)

Yue calculated the radius of a proton and neutron with constants that are similar, but not equal to the constants presented in CODATA 2014. Yue's method is adjusted for the constants listed in the CODATA Internationally recommended 2014 values of the Fundamental Physical Constants, Physical Measurement Laboratory of National Institute of Standards and Technology.[38] Yue's Mass Neutron (Mn): 1.674954386 E-27 kg, while CODATA 2014: 1.674927471 E-27 kg. Yue's Mass Proton (Mp): 1.672648586 E-27 kg, while CODATA 2014: 1.672621898 E-27 kg. Yue's Electric charge of a proton (Qp): 1.602189246 E-19 C, while CODATA 2014: 1.6021766208(98) E-19 C. Yue's Electric charge of an electron (Qe): 1.602189246 E-19 C, while CODATA 2014: 1.6021766208(98) E-19 C.

Total mass for releasing neutron disintegration is Mn – Mp using CODATA values:

Mn – Mp = 1.674927471 E-27 kg – 1.672621898 E-28 kg = 2.305573   (4)
E-30 kg

Radius of a neutron calculated as:

$$Rn = K * Qp * Qe / ((Mn - Mp) * C^2) \tag{5}$$

$$Rn = \frac{8.987551786262* E9\frac{m}{F} * 1.602189246* E-19\ C* 1.602189246*E-19\ C}{2.305573* E-30\ kg* (2.99792458*E10\ m/s)2} \tag{6}$$

Rn = 1.113393668161 E-15 m (calculated with CODATA 2014 (7) constants, ref 38)

$$Rp = (Mp/Mn)^{1/3} \times Rn \tag{8}$$

Rp = (1.67262189 E-27 kg/1.67492747 E-27 kg)$^{1/3}$ * 1.11339366816 (9) E-15m

Rp = 1.11288256362 E-15 m (calculated with CODATA 2014 (10) constants, ref 38)

The radius of a neutron calculated with CODATA 2014 constants is equal to 1.113393668161 E-15 m, and the radius of a proton calculated with CODAT 2014 constants is equal to 1.11288256362 E-15 m.

Solving for mass and diameter of a quadsitron is as follows:

STEP #1: Solve for mass of one neutrino

It is deduced that a neutrino is comprised of two quadsitrons. The two quadsitrons are configured to trap energy between three of the four poles of each of the two quadsitrons. The mass of a single quadsitron is half of the mass of a neutrino.

Energy = mass multiplied by speed of light squared = m * c^2 (11)

Mass of three neutrinos = 0.32 ± 0.081 eV/c^2 [ref 53] (12)

Approximate mass of one neutrino = 0.32 ± 0.081 eV/c^2 / 3 = (13) 0.1066 eV/c^2

$1 \text{ eV}/c^2 = 1.782661907 \text{ E-36 kg [ref 38]}$ (14)

Mass (kilograms) of 3 neutrinos = $(0.32 \text{ eV}/c^2 * 1.782661907 \text{ E-36}$ (15)
kg/ eV/c²)

Mass (kilograms) of 3 neutrinos = 0.57045181024 E-36 kg   (16)

Mass (kilograms) of 1 neutrino or 1 anti-neutrino = 0.57045181024   (17)
E-36 kg / 3

Mass (kilograms) of 1 neutrino or 1 anti-neutrino = 1.90150603413   (18)
E-37 kg.

STEP #2: Solve for mass of quadsitron (Mq)

It is theorized that two quadsitrons make up a neutrino or anti-neutrino. Two quadsitrons combine to trap energy as seen in figure 2.

Anti-neutrino mass (kg) = Mve = 1.90150603413 E-37 kg   (18)

Mq is ½ the size of an anti-neutrino = 1.90150603413 E-37 kg / 2   (19)

The mass of one quadsitron is equal to 9.5075301706 E-38 kg.   (20)

STEP #3: Solve for number of quadsitrons comprising a neutron designated variable 'a'

Knowing the mass of a single quadsitron, and utilizing the mass of a known neutron, the number of quadsitrons present in a neutron can be calculated.

Mass of a neutron = Mn = 1.6749.27471 E-27 kg [ref 38]   (21)

Mass of a quadsitron = Mq = 9.5075301706 E-38 kg (derived from (20)
STEP #2)

$$Mn = a * Mq \tag{22}$$

$$a = Mn / Mq \tag{23}$$

$$a = 1.674927471 \text{ E-27 kg} / 9.5075301706 \text{ E-38 kg} \tag{24}$$

$$a = 1.7616851495 \text{ E10} \tag{25}$$

Number of quadsitrons comprising a neutron using Mn and Mq is 1.7616851495 E10.

STEP #4: Solve for number of quadsitrons comprising a proton designated variable 'b'

Knowing the mass of a single quadsitron, and utilizing the known mass of a proton, the number of quadsitrons present in a proton can be calculated.

$$\text{Mass of a proton} = Mp = 1.672621898 \text{ E-27 kg [ref 38]} \tag{26}$$

$$\text{Mass of a quadsitron} = Mq = 9.5075301706 \text{ E-38 kg (derived from} \tag{20}$$
STEP #2)

$$Mp = b * Mq \tag{27}$$

$$b = Mp / Mq \tag{28}$$

$$b = 1.672621898 \text{ E-27 kg} / 9.5075301706 \text{ E-38 kg} \tag{29}$$

$$b = 1.7592601527 \text{ E10} \tag{30}$$

Number of quadsitrons comprising a proton using Mp and Mq is 1.7592601527 E10.

STEP #5: Solve for neutron volume:

The radius of a neutron is taken from the calculated value using CODATA 2014 constants as Rn = 1.113393668161 E-15 m Eq. (7). Approximating that a neutron is in the shape of a sphere, utilizing the radius of a neutron, the volume of a neutron can be calculated.

Neutron volume = Vn = 4/3 * π* $(Rn)^3$ (31)

Radius of a neutron = Rn = 1.113393668161 E-15 m   (7)

Vn = 1.333333333 * 3.141592653 * $(1.113393668161 \text{ E-15 m})^3$ (32)

Vn = 4.188790202 * 1.380213406264 E-45 m^3 (33)

Vn = 5.781424392826 E-45 m^3 (34)

The calculated volume of a neutron is 5.781424392826 E-45 m^3.

STEP #6: Solve for proton volume:

The radius of a proton is taken from the calculated value using CODATA 2014 constants as Rp being equal to 1.11288256362 E-15 m Eq. (10). Approximating that a proton is in the shape of a sphere, utilizing the radius of a proton, the volume of a proton can be calculated.

Proton volume = Vp = b * Vq = 4/3 * π* $(Rp)^3$ (35)

Radius of a proton = Rp = 1.11288256362 E-15 m   (10)

Vp = 1.333333333 * 3.141592653 * $(1.11288256362 \text{ E-15 m})^3$ (36)

Vp = 4.188790202 * 1.378313513406 E-45 m^3 (37)

Vp = 5.773466140241 E-45 m^3 (38)

The calculated volume of a proton is 5.773466140241 E-45 m³.

STEP #7: Solve for change in neutron volume to proton volume

If a neutron is approximated to be a sphere and a proton is a sphere, and if the neutron is larger by volume than a proton, the change in volume of a neutron during Beta decay to a proton can be calculated.

Change in Neutron Volume to Proton Volume = $V_n - V_p$ (39)

Using radius of a neutron being 1.113393668161 E-15 and radius of a proton being 1.11288256362 E-15 the change in volume of a neutron decaying to a proton is calculated:

$V_n - V_p$ = 5.773466140241 E-45 m³ - 5.781424392826 E-45 m³   (40)

$V_n - V_p$ = 7.958252585 E-48 m³   (41)

Change in volume of a free neutron to volume of a proton is 7.958252585 E-48 m³.

STEP #8: Solve for diameter of a quadsitron (Dq) by means of volume of neutron (Vn):

A neutron is approximated to be in the shape of a sphere. A quadsitron has four poles with each pole separated by 120 degrees equiradius from the center of the quadsitron. Given the four poles are equiradius from the center of the quadsitron, the diameter of a quadsitron can be approximated by treating the quadsitron as if it were a sphere. A quadsitron may be likened to a sphere with four poles. Treating the quadsitron as a sphere allows for the calculation of the gross approximate volume of a quadsitron. Knowing the number of quadsitrons present in a neutron as calculated by weight allows to solve for the diameter of a quadsitron.

$V_n = a * V_q$ (42)

$$a = Vn / Vq = Mn / Mq = 1.7616851495 \text{ E}10 \tag{43}$$

$$Vq = Vn / a \tag{44}$$

$$Vq = 4/3 * \pi * (Rq)^3 \tag{45}$$

$$Vq = 4/3 * \pi * (Dq/2)^3 \tag{46}$$

$$Dq = (\sqrt[3]{(Vq * \tfrac{3}{4})/\pi}) * 2 \tag{47}$$

$$Dq = (\sqrt[3]{((\tfrac{Vn}{a}) * \tfrac{3}{4})/\pi}) * 2 \tag{48}$$

$$Dq = (\sqrt[3]{\frac{\tfrac{4}{3}* \pi*(Rn)^\wedge 3 *\tfrac{3}{4}}{\pi * a}}) * 2 \tag{49}$$

$$Dq = (\sqrt[3]{(Rn)^\wedge 3/a}) * 2 \tag{50}$$

$$a = 1.7616851495 \text{ E}10 \tag{51}$$

$Rn = 1.113393668161$ E-15 m (calculated with CODATA 2014 (7) constants, ref 38)

$$Dq = (\sqrt[3]{(Rn)^\wedge 3/1.7616851495 \text{ E}10}) * 2 \tag{52}$$

$$Dq = (\sqrt[3]{(1.113393668161E - 15m)^\wedge 3/1.7616851495 \text{ E}10}) * 2 \tag{53}$$

$$Dq = (\sqrt[3]{1.380213406264E - 45m/1.7616851495 \text{ E}10}) * 2 \tag{54}$$

$$Dq = (\sqrt[3]{78.3462020245 \text{ E} - 57\,m}) * 2 \tag{55}$$

$$Dq = (4.278970732943 \text{ E-}19 \text{ m}) * 2 \tag{56}$$

$$Dq = 8.557941465886 \text{ E-}19 \text{ m} \tag{57}$$

The diameter of a quadsitron using Vn is 8.557941465886 E-19 m.

$Rq = Dq/2 = 8.557941465886 \text{ E-19 m} / 2$ (58)

$Rq = 4.278970732943 \text{ E-19 m}$ (59)

The radius of a quadsitron using the volume of a neutron (Vn) is 4.278970732943 E-19 m.

STEP #9: Solve for diameter of a quadsitron (Dq) by means of volume of a proton (Vp)

A proton is approximated to be in the shape of a sphere. A quadsitron has four poles with each pole separated by 120 degrees equiradius from the center of the quadsitron. Given the four poles are equiradius from the center of the quadsitron, the diameter of a quadsitron can be approximated by treating the quadsitron as if it were a sphere. Treating the quadsitron as a sphere allows for the calculation of the gross approximate volume of a quadsitron. Knowing the number of quadsitrons present in a proton as calculated by weight allows to solve for the diameter of a quadsitron.

$Vp = b * Vq$ (60)

$b = Vp / Vq = Mp / Mq = 1.7592601527 \text{ E10}$ (61)

$Vq = Vp / b$ (62)

$Vq = 4/3 * \pi * (Rq)^3$ (63)

$Vq = 4/3 * \pi * (Dq/2)^3$ (64)

$Dq = (\sqrt[3]{(Vq * \frac{3}{4})/\pi}) * 2$ (65)

$Dq = (\sqrt[3]{((\frac{Vp}{b}) * \frac{3}{4})/\pi}) * 2$ (66)

$Vp = 4/3 * \pi * (Rp)^3$ (67)

$$Dq = \left(\sqrt[3]{\dfrac{\frac{4}{3} * \pi * (Rp)^{\wedge}3 * \frac{3}{4}}{\pi * b}} \right) * 2 \tag{68}$$

$$Dq = \left(\sqrt[3]{(Rp)^{\wedge}3/b} \right) * 2 \tag{69}$$

b = 1.7592601527 E10   (61)

Rp = 1.11288256362 E-15 m (calculated with CODATA 2014   (10) constants, ref 38)

$(Rp)^3$ = 1.378313513406 E-45 m³   (70)

$$Dq = \left(\sqrt[3]{(Rp)^{\wedge}3/1.7592601527 \text{ E10}} \right) * 2 \tag{71}$$

$$Dq = \left(\sqrt[3]{(1.11288256362 \text{ E} - 15 \text{ m})^{\wedge}3/1.7592601527 \text{ E10}} \right) * 2 \tag{72}$$

$$Dq = \left(\sqrt[3]{1.378313513406E - 45m/1.7592601527E10} \right) * 2 \tag{73}$$

$$Dq = \left(\sqrt[3]{78.3462020264949E - 57 \text{ m}} \right) * 2 \tag{74}$$

Dq = (4.2789707329791 E-19 m) * 2   (75)

Dq = 8.557941465958 E-19 m   (76)

The diameter of a quadsitron using volume of a proton (Vp) is 8.557941465958 E-19 m.

Rq = Dq/2 = 8.557941465958 E-19 m / 2   (77)

Rq = 4.2789707329791 E-19 m   (78)

The radius of a quadsitron using Vp is 4.278970732979 E-19 m.
The diameter of a quadsitron using Vn is 8.557941465886 E-19 m.
The radius of a quadsitron using Vn is 4.278970732943 E-19 m.
Calculated radius of a quadsitron per Vn and Vp (STEP #8 and STEP #9) are equal to 4.2789707329 E-19 m.

Diameter of quadsitron per Vn & Vp (STEP #8 & STEP #9)
8.557941465 E-19 m. (79)

STEP #10: Solving for energy anti-neutrino (VE) in Beta decay of free neutron using Mq

The known resultant kinetic energy of an electron in the Beta decay of a free neutron is 0.51 MeV. The known kinetic energy of an anti-neutrino in the Beta decay of a free neutron is 1.29 MeV - 0.51 MeV to equal 0.78 MeV.[54] Utilizing the weight of a quadsitron, the kinetic energy associated with an anti-neutrino should be calculable. To do this, first the number of quadsitrons associated with an electron is determined by taking the weight of the electron and dividing by the weight of a single quadsitron. Then the number of quadsitrons associated with the change in weight of a neutron decaying to a proton is determined by taking the weight associated with the change in weight of a neutron to a proton and dividing this number by the weight of a single quadsitron. Taking the number of quadsitrons associated with the change in weight of a neutron to a proton and subtracting the number of quadsitrons associated with an electron results in the number of quadsitrons associated with the energy of the anti-neutrino in the Beta decay of a free neutron.

Beta decay of a free neutron is generally recognized to be 1.29 MeV = 0.51 MeV + 0.78 MeV; with the known energy of an electron = 0.51 MeV. CODATA 2014 provides neutron-proton mass difference energy equivalent in MeV as $(M_n - M_p) * c^2$ as 1.29333205 MeV.[38] The CODATA 2014 electron mass energy equivalent $M_e * c^2$ in MeV is 0.5109989461 MeV.[38] Therefore, utilizing CODATA 2014 values:

Beta decay = 1.29333205 MeV = 0.5109989461 MeV + kinetic (80)
energy of a neutrino

Kinetic energy of a neutrino = 1.29333205 MeV - 0.5109989461 MeV (81)

Kinetic energy of a neutrino = 0.782333104 MeV (82)

Me = g * Mq, where g is the number of quadsitrons comprising an electron (83)

Mq = 9.5075301706 E-38 kg (derived from STEP #2) (20)

Me = 9.10938356 E-31 kg [ref 38] (84)

g = Me / Mq (85)

g = 9.10938356 E-31 kg / 9.50753017 E-38 kg (86)

g = 9.581230 E6 (87)

Change in mass Mn to Mp = 2.30557377 E-30 kg [ref 38] (88)

y = Number quadsitrons (Mn - Mp) / Mq = 2.30557377 E-30 kg / 9.50753017 E-38 kg (89)

y = Number quadsitrons in (Mn - Mp) / Mq = 2.4249975 E7 (90)

z = Number of quadsitrons following removal of electron = y – g (91)

z = Number of quadsitrons removal of electron = 2.4249975 E7 - 9.581230 E6 (92)

z = Number of quadsitrons following removal of electron = 1.4668745 E7 (93)

Energy of anti-neutrino in Beta decay = VE = z * Mq (94)

VE = z * Mq (95)

VE = 1.4668745 E7 * 9.5075301706 E-38 kg   (96)

VE = 1.3946353565 E-30 kg   (97)

mass = 1.782661907 E-36 kg / 1 eV/c^2 [ref 38]   (98)

Energy result of Beta decay = Ve / mass conversion   (99)

Energy of Beta decay = 1.3946353565 E-30 kg / 1.782661907 E-36   (100)
kg / 1 eV/c^2

Energy result of Beta decay = 0.782333066 E6 eV/c^2 (101)

The calculated energy result of Beta decay by using Mq is 0.7823330666 MeV/c^2.

Energy result of Beta decay by using Mq is 0.7823330666 MeV/c^2, which is similar to the published measured value assigned to the anti-neutrino of the Beta decay of a free neutron which is 0.78 MeV/c^2 or even closer to the above calculated kinetic energy of a neutrino = 0.782333104 MeV/c^2 using values for energy associated with Beta decay and the electron acquired from CODATA 2014, reference 38.

STEP #11: Solve for number of quadsitrons comprising a neutron designated variable 'a'

Neutron Volume = Vn = 5.781424392826 E-45 m^3 (34)

Quadsitron volume = Vq = 4/3 * π * (Dq/2)^3 (102)

Vq = 4/3 * π * (8.557941465 E-19 m/2)^3 (103)

Vq = 4/3 * 3.141592653 * (4.2789707325 E-19 m)^3 (104)

Vq = 3.281755803459 E-55 m^3 (105)

$a = Vn / Vq = 5.781424392826 \text{ E-45 m}^3 / 3.281755803459 \text{ E-55 m}^3$ (106)

$a = 1.7616863469038 \text{ E10}$ (107)

The volume of a quadsitron if the shape is treated as a sphere is $3.281755803459 \text{ E-55 m}^3$.

The number of quadsitrons present in volume of neutron per Vn / Vq = $1.7616863469038 \text{ E10}$.

The number of quadsitrons comprising a neutron using Mn and Mq (as shown in STEP #3) is 1.7616851494 E10.

The number of quadsitrons calculated by volume and calculated   (108) by weight comprising a neutron are both equal at 1.76168 E10.

STEP #12: Solve for number of quadsitrons comprising a proton designated variable 'b'

Volume of a proton = Vp = $5.773466140241 \text{ E-45 m}^3$ (38)

Volume of a quadsitron Vq = $3.281755803459 \text{ E-55 m}^3$ (105)

Proton Volume = Vp = b * Vq   (109)

b = Vp / Vq   (110)

$b = 5.771760500361 \text{ E-45 m}^3 / 3.280788888658 \text{ E-55 m}^3$ (111)

$b = 1.7592599513 \text{ E10}$ (112)

The number of quadsitrons present in volume of proton per Vp / Vq = 1.7592599513 E10.

The number of quadsitrons comprising a proton using Mp and Mq (as shown in STEP #4) is 1.7592601527 E10.

The number of quadsitrons calculated by volume and calculated (113) by weight for a proton are both equal at 1.75926 E10.

STEP #13: Solve for number quadsitrons in change in neutron volume to proton volume

The number of quadsitrons associated with the change in mass when a free neutron undergoes Beta decay to a proton was calculated previously. Here the number of quadsitrons associated with the change in volume when a free neutron undergoes Beta decay to a proton is calculated using the derived diameter of a quadsitron. The number of quadsitrons associated with change of a free neutron to a proton during Beta decay calculated by change in weight should equal the number of quadsitrons associated with change of a free neutron to a proton during Beta decay calculated by volume.

Neutron Volume = Vn = 5.781424392826 E-45 m^3 (as derived in (34) STEP #5)

Proton Volume = Vp = 5.773466140241 E-45 m^3 (as derived in (38) STEP #6)

Quadsitron volume = Vq = 3.281755803459 E-55 m^3 (as derived (105) in STEP #11)

Change in Neutron Volume to Proton Volume = Vn – Vp (114)

Vn - Vp = 5.781424392826 E-45 m^3 - 5.773466140241 E-45 m^3 (115)

Vn – Vp = 7.958252585 E-48 m^3 (116)

Number of quadsitrons in change in volume of Neutron to Proton (117) = (Vn – Vp) / Vq

$y = (V_n - V_p) / V_q = 7.958252585 \text{ E-48 m}^3 / 3.281755803459$ (118)
E-55 m^3

$y = (V_n - V_p) / V_q = 2.42499840 \text{ E7}$ (119)

The number of quadsitrons comprising the resultant change in volume of a free neutron per Beta decay to a proton is 2.42499840 E7.

Neutron mass (weight) = M_n = 1.6749.27471 E-27 kg [ref 38] (120)

Proton mass (weight) = M_p = 1.6726.21898 E-27 kg [ref 38] (121)

Quadsitron mass (weight) = M_q = 9.5075301706 E-38 kg (as (20) derived in STEP #2)

Change in Neutron mass to Proton mass in Beta decay = M_n − (122)
$M_p = y * M_q$

$y = (M_n - M_p) / M_q$ (123)

$y = (1.674927471 \text{ E-27 kg} - 1.672621898 \text{ E-27 kg}) / 9.5075301706$ (124)
E-38 kg

$y = 2.305573 \text{ E-30 kg} / 9.5075301706 \text{ E-38 kg}$ (125)

$y = 2.4249968 \text{ E7}$ (126)

The number of quadsitrons calculated by volume and calculated (127) by weight for the change of a neutron to a proton is both approximately equal at 2.42499 E7.

STEP #14: Solve for change in radius of a neutron to proton

During Beta decay the neutron transforms into a proton and the diameter changes.

Rn = 1.113393668161 E-15 m (calculated with CODATA 2014 (7) constants, ref 38)

Rp = 1.11288256362 E-15 m (calculated with CODATA 2014 (10) constants, ref 38)

Change in Radius of a Neutron to Proton = Rn – Rp (128)

Rn – Rp = 1.113393668161 E-15 m - 1.11288256362 E-15 m (129)

Rn – Rp = 5.11104551 E-19 m (130)

The change in the radius of a neutron to a proton is 5.11104551 (131) E-19 m.

STEP #15: Solve for number of quadsitrons in the radius of a proton

During Beta decay the free neutron transforms into a proton and the diameter changes. A number of quadsitrons comprise a neutron and proton, therefore number of quadsitrons comprising a proton can be calculated.

Dq per Vn and Vp equals to 8.557941465 E-19 m (STEP #8 and #9) (79)

Proton radius = t * Dq (132)

Rp = t * Dq (133)

Number quadsitrons in radius of proton = t = Rp/Dq (134)

$$t = Rp/Dq = 11,128.8256362 \text{ E-19 m}/8.557941465 \text{ E-19 m} \tag{135}$$

$$t = Rp/Dq = 1,300.409179 \tag{136}$$

Number of quadsitrons comprising radius of a proton is (137) 1,300.409179; truncated is 1,300.

STEP #16: Solve for number of neutrinos in the radius of a proton

Two quadsitrons comprise a neutrino.

Neutrinos comprising the radius of a proton = t/2 (138)

$$t/2 = 1300/2 \tag{139}$$

$$t/2 = 650 \tag{140}$$

The number of neutrinos in the radius of a proton is 650. (141)

STEP #17: Solve for the physical difference between a neutron and a proton as a result of Beta decay.

Both neutron and a proton are theorized to exist as a sphere.

Calculated diameter of a quadsitron per Vn and Vp (STEP #8 and STEP #9) is equal to 8.557941465 E-19 m.

The number of quadsitrons calculated by volume and calculated by weight for a neutron are both equal at a = 1.76168 E10 (derived in STEP #11).

The number of quadsitrons calculated by volume and calculated by weight for a proton are both equal at b = 1.75926 E10 (derived in STEP #12).

The number of quadsitrons calculated by volume and calculated by weight for the change of a neutron to a proton are both approximately equal at y = 2.42499 E7 (derived in STEP #13).

The change in the radius of a neutron to a proton is 5.11104551 E-19 m (derived in STEP #14).

A proton is theorized to be sphere with a radius of 1300 (142)
quadsitrons.

A neutron is theorized to be a sphere with a radius of 1301 (143)
quadsitrons.

This difference in the radius of a neutron and the radius of a proton
is smaller than the diameter of a single quadsitron.

Percent change neutron to proton versus quadsitron = (Rn-Rp) / (144)
Dq * 100%

(Rn-Rp) / Dq * 100% = 5.11104551 E-19 m / 8.557941465 E-19 m (145)
* 100%

(Rn-Rp) / Dq * 100% = 59.7228% (146)

Percent change of a neutron to a proton versus diameter of a (147)
quadsitron is 59.7228%

The percent change of a neutron to a proton versus diameter of
a quadsitron is 59.7228% or almost sixty percent of the diameter of a
quadsitron. The fact that the superficial layer of a neutron has a thickness
of 5.11104551 E-19 m suggests that there is a surface layer comprised of
a single layer of quadsitrons which covers the surface of a proton to
generate a neutron. The quadsitrons comprising the superficial layer
are seated on the surface of the proton, with the positive pole of the
quadsitrons comprising the surface of the proton embedded forty percent
inside the quadsitrons comprising the superficial layer of the resultant
neutron. The superficial layer of quadsitrons covering the surface of the
proton point their neutral pole outward, creating a neutron. Thus, a free
neutron releases 1.29 MeV of energy while shedding a superficial layer of
quadsitrons to expose a proton which emerges from inside the neutron.

LIST OF VARIABLES

a = number of quadsitrons comprising a neutron by weight and volume
a0 = Bohr radius
b = number of quadsitrons comprising a proton by weight and volume
bne is neutron-electron scattering length
c = speed of light
e = electron
fm = fentimeter
g = number of quadsitrons comprising an electron by weight and volume
h = number of quadsitrons comprising an anti-neutrino
m = number of quadsitrons comprising the diameter of a neutron
n = neutron
q = quadsitron
p = proton
s = number of quadsitrons comprising the diameter of a proton
t = number of quadsitrons comprising the radius of a proton
w = number of quadsitrons comprising the diameter of a neutrino or an anti-neutrino
y = number of quadsitrons comprising the change from neutron to proton
z = number of quadsitrons comprising the change of neutron to proton minus electron
Dn = diameter of a neutron
Dq = diameter of a quadsitron
Dp = diameter of a proton
Dve = diameter of an anti-neutrino or neutrino
K = Electromagnetic constant
Mn = mass of neutron
Mp = mass of proton
Me = mass of electron
Mve = mass of anti-neutrino or neutrino

Mq = mass of quadsitron

Qp is electric charge of a proton

Qe is the electric charge of an electron

Rn = radius of a neutron

Rp = radius of a proton

Rq = radius of a quadsitron

Rve = radius of an anti-neutrino or neutrino

Vn = volume of neutron

Vq = volume of a quadsitron

Vp = volume of a proton

Ve = volume of a neutrino

VE = Energy of anti-neutrino in Beta Decay

TABLE OF KNOWN CONSTANTS/QUANTITIES

Bohr radius = a0 = 5.2917721067 E-11 m [ref 38]

Change in the radius of Rn to Rp = 5.11096351 E-19 m [ref 149]

Change in the diameter of Dn to Dp = 10.22186352 E-19 m [ref 149]

Electric charge of an electron (Qe) 1.6021766208(98) E-19 C [ref 38]

Electric charge of a proton (Qp) 1.6021766208(98) x E-19 C [ref 38]

Electromagnetic constant = K= 8.987551786262 E+9 m/F [ref 149]

Electron Mass (weight) = Me = 9.10938356 E-31 kg [ref 38]

Electron mass energy equivalent $m_e c^2$ in MeV is 0.5109989461 MeV. [ref 38]

Electron radius (classical) = Re = 2.81794 E-15 m [ref 71]

1 electron volt = 1 eV = 1.6021766208 E-19 joules [ref 38]

$1\ eV/c^2$ = 1.782661907 E-36 kg [ref 38]

fentimeter = fm = E-15 m [ref 24]

Free neutron root mean square radius = 0.8269 E-15 m [ref 156]

Free proton root mean charge square radius = 0.8305 fm [ref 156]

Kinetic energy of an electron = 0.51 MeV [ref 38]

Mass of three neutrinos = 0.32 ± 0.081 eV [ref 53]

Maximum kinetic energy Beta decay of free Neutron = 1.29 MeV-0.51 MeV = 0.78 MeV [ref 54]

Neutron Mass (weight) = Mn = 1.674927471 E-28 kg [ref 38]

Neutron-Proton mass difference is 1.29 MeV [ref 53]

Neutron-Proton mass difference energy equivalent in MeV as (Mn-Mp) c^2 as 1.29333205 MeV. [ref 38]

Neutron-Proton mass difference (Mn - Mp) = 2.30557377 E-30 kg [ref 38]

Neutron radius = Rn = 1.113284057367 E-15 m [ref 149]

pi = π = 3.141592653 [ref 157]

Planck constant = 6.626070040 E-34 joule-second [ref 38]

Planck constant = 6.62607015 E-34 joule-second, updated 2018. [ref 52]

Proton Mass (weight) = Mp = 1.672621898 E-28 kg [ref 38]

Proton radius = Rp = 1.112772961016 E-15 m [ref 149]

Proton root mean square radius = Mainz electron scattering = 0.8775 E-15 m [ref 145]

Proton root mean square radius = hydrogen 2S-2P lamb shift = 0.84087 E-15 m [ref 147]

Speed of light = c = 2.99792458 E+8 m/s [ref 38]

PULLING BACK THE WIZARD'S CURTAIN TO PEER BEYOND

LIFE GOES ON DAY by day. The stabbing pain of the morning alarm clocks erupts in your head, like a semi-trailer truck unexpectedly crashing into a brick wall, wrenching you from the tranquil daze of a deep sleep. One wakes up with the sun on the horizon to initiate the venture off work, whatever that role in the social fabric might be. You take a shower to reset the sensors which cover the body setting their baseline to neutral. You prepare breakfast, then listen to the morning news program. You tell yourself you are listening to the news to get the weather report or the traffic report, but actually deep inside your psychic you are hunting to see if the world is still in one piece. Residing in an industrialized nation, you are searching to see if a crazed fanatical leader hadn't thrown the world into an epic global crisis from which one's fabricated plastic life might not recover.

Actually, what many are hoping is for someone to pull back the curtain and show them the workings of the real world. Most of us likely do not venture to see behind the curtain due to the hurriedness of daily life in the twenty-first century keeping most of the population too busy to have the time to contemplate an alternative life; in addition, life is generally bearable just as it is, so why question it. Some hope for the curtain to fall to the floor, to expedite facing reality.

The truth of the matter is that the Earth orbits the Sun as the third planet in the solar system. Our solar system resides in the Orion Spur of the Milky Way Galaxy about 40 million light years from the swirling central vortex of the galaxy. Within ten light years from our sun are eleven other stars. The nearest fiery celestial bodies closest to our planet is a collection of three stars generally referred to as Alpha Centuri. The closest star to our sun, approximately 4.2 light years distance is known as Alpha Centuri proixima. A little further distant is Alpha Centuri B orbiting around Alpha Centuri A. There are then two double star systems and four small red dwarf stars with in the ten light year radius from Earth. Travel at one tenth the speed of light, which would seem to be a practical goal for space travel, would suggest that even at a speed of one tenth the speed of light, it would take one hundred years to reach the edge of this arbitrary ten light year boundary.

The speed of light is roughly 720 million miles an hour. Therefore, one tenth the speed of light is 72 million miles an hour. The fastest speed a human constructed object has attained is a recent satellite launched into the deep reaches of the solar system traveling at 36 thousand miles an hour. Our greatest achievement in space travel is a velocity 200 times less than one tenth the speed of light. Within ten light year from Earth or at best a travel time of 20,000 years to reach the edge of the boundary, there is no sun that humans would wish to migrate to in an effort to expand the human race or reach out to in order to save the human race from extinction.

The reality that haunts us from behind the curtain is that despite the progress we humans have achieved, the human race is destined to fade from the eyes of the universe unless we are able to accurately solve the mysteries of light, magnetism and gravity.

As the curtain gets thrust aside, what if knowledge of the workings behind the curtain revealed that all which surrounds us is not the result of random events, but instead the makings of our environment deeply rooted in absolute purpose. Science has led the populace to believe that our whole world has teetered on a constellation of random happenings.

We have been taught that the very existence of humanity is due to Evolution. This process, which occurred billions of years ago owing its initiation to a pool of water teaming with essential atoms, which was struck by a lightning bolt, which then resulted in organization of random atoms into molecules and then into the building blocks of the RNA and DNA. The essence of Evolution is that enough time passed that a sufficient number of random events occurred in the proper succession to result in organization and to create life on planet Earth. The construct of music is random. We have been told that when we fall in love, this is the ultimate act of serendipity. Are these concepts all based on random chance; or is there an underlying driving force discretely organizing the progress of life.

Dissecting the programming code of the Human Immunodeficiency Virus (HIV) genome reveals that the construct of the 4915 nucleotides comprising the virus's genome is astronomically impossible for the HIV genome to be the result of random events. Eight individual segments of the HIV genome are overlapping. One segment of the HIV genome is used three differing times. The HIV genome is an elegant example of a base-three codon code interweaved into a base-four nucleotide genetic code. Eight segments of the HIV genome take advantage of frame-shifting of the base-three codon code to generate a data compression technique which is far superior to any digital computer data compression that has been crafted by humans.

The universe we live in is built on the foundation of the coordinated interactions of quadsitrons and energy. The crafters of the human race knew the physics down to the level of the quadsitron. The designers of life worked from the atomic level all the way through the macro level of multicellular organism to the ultimate macro level of designing an interactive balanced ecosystem that could flourish across the surface of a planet such as the Earth. The designers had to take into account the Earth started as a raw planet scorched by the radiation of the sun, fiery and arid. That in the beginning, the atmosphere would be dominated by helium and hydrogen, the surface dotted with active volcanoes spewing a toxic mixture of carbon dioxide, carbon dioxide, and sulfuric acid. The designers crafted a biologic world from an essence of understanding the interactions of the atomic structures of base atoms, interfusing the chemistry to create biologic systems collected in a unit cell. Diversifying the design of the unit cell led to an expansion of life from bacteria to multi-cellular organisms until the human body was practical.

Hidden behind the curtain, is the Prime Genome program, responsible for all of life on the planet and the construct of the ecosystems which flourish across the planet. The objectives of the Prime Genome are to survive, given the prevailing environmental factors and to mold the biologic environment on the Earth to the point to support an intelligent life form sufficient to understand its own existence and to reconstruct the program in a manner that the program could be launched out into space to populate another planet either by travel by the intelligent life or by seeding the universe with copies of the Prime Genome.

Those living in third world countries, where daily conditions are harsh at best, understand that survival may be an everyday challenge, and somedays many do not survive. Often in the industrialized world life appears more certain than uncertain. The Internet, modern medicine, emergency services which can be activated with a phone call, combustion or electric means of transportation, fast food restaurants of all kinds, and the abundance of twenty-four-hour convenience stores all help prop up the facade that with enough money, everyday life can be luxurious. Reality is that the human existence is locked in a deadly struggle to survive in order to maintain its place in the universe.

The truth may be that the Prime Genome may be responsible for having created all of the life that has ever existed on the planet. The Prime Genome may have constructed the human brain with a fixed set of predetermined memories. We don't question when an animal exhibits intelligence in seeking food, mating or rearing offspring; we blindly interpret it as 'natural instinct' instead of recognizing and accepting the fact that the animal is exhibiting elements of pre-set programming. Humans may indeed have elements of preprogramming hardwired into the neuro network in their brain. Music may tap into the pre-wired files in the brain, not only taping into emotion, but also unlocking files. The selection of a mate in some may be satisfying pre-wired conditions for a mate that is present at birth. Our existence, and our mission as a species, may be hardwired into our brains, a need to survive and flourish, dominating the planet's surface, controlling the earth's resources to ensure that the Prime Genome survives whatever catastrophe may paralyze the Earth and, if possible, spread the Prime Genome further into the galaxy.

THE HUMAN SUPER POWER: UNLOCKING THE DARK FILES OF THE HOMO SAPIENS' BRAIN

Introduction

THE ANATOMY OF THE human brain has been known for centuries, but the inner workings of the human brain remains one of Nature's most coveted secrets. Yet, to understand how the human brain functions, provides access to the most valuable tools known to humankind.

Sigmund Freud (1856–1939) provided medical science with many of the initial founding principles for how the human brain functioned. He described the human brain as having three terms for mental phenomena, those being 'conscious', 'preconscious' and 'unconscious'. Freud further identified three realms, regions, provinces which were divisions of a person's mental apparatus to include 'self', 'superego' and 'id'.[158] See Figure 208. 'Self' was in essence representative of an individual's perception of one's existence. 'Superego' essentially represented morality. 'Id' represented the decisions that leaned more toward pleasure and survival, and frequently challenged the sense of morality. The province of 'id' often being associated with one's libido.

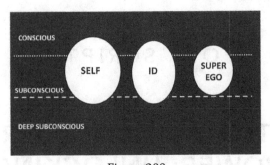

Figure 208

Self, ID, Super Ego.

This text in no way wishes to unseat Sigmund Freud's work regarding its profound reputation as being genius regarding his description of the psychology of the human brain and human behavior. Sigmund Freud's work has been the cornerstone of Medicine's approach to understanding the human psyche and intervening when human behavior has bordered on illness or has become pathological. Sigmund Freud's tireless efforts in detailing human psychology can only be amplified.

SELF, MORALITY, SURVIVALIST, CONCIERGE, HOTEL MANAGER

Expanding the base knowledge with parallel advancements in our understanding of computers, human anatomy, and behavior/brain function, there are likely at least five entities comprising human behavior, to include 'self', 'morality', 'survival', 'hotel manager' and 'concierge'. See Figure 209.

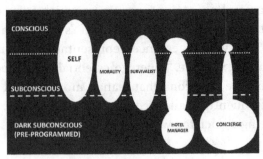

Figure 209

Self, Morality, Survivalist, Hotel Manager, Concierge.

Not to be confused with the listed modules operating in the human brain to include self, survivalist, morality, hotel manager, and concierge, there are three levels of consciousness. The three levels of consciousness include the 'conscious' state, 'subconscious' state, and the 'dark subconsciousness'. Consciousness is the act of being aware of one's surroundings in real time. The subconscious is the operations of the brain which are not functions which actively fill the awareness of a human.

The subconscious acts much like a memory storage device, which catalogues memories and keeps them offline in a retrievable file system until needed by the conscious brain. Most persons actively interface with the subconscious state of the brain when taking a trivia test requiring recall of distant, sometimes obscure memories, which are generally memories a person has acquired, just not immediately available at the conscious level. Depending upon the person's inherent speed of memory file access and how important the fact was when memorized often dictates how long it takes for an individual to remember a fact when asked for such information.

The third level of consciousness is the 'dark subconscious'. The dark subconscious refers to the portion of the brain which contains pre-programmed memory files. Pre-programmed memory files assist human in functioning throughout life and guide mankind in progress.

SELF

The self is the conscious knowing of one's existence in the universe and relation to the surrounding environment. Self, generally operates predominantly in the conscious portion of the brain, but is able to interact with the subconscious and dark subconscious. See Figure 210. Self is able to perceive and act on information input from exterior sensors scattered about the skin and internal organs of the body. Self has a loadable memory. Self actively interacts with short-term and long-term memory. Being able to store important memories allows 'self' to remember facts which lead to individualization of the person such as memories of family members, memories of travel, memories of achievements and failures and memories of facts necessary for a person to be functional in the society they reside in throughout their life. Self also has access to the dark subconscious and a certain personalized skill set, which is related to an individual's talent for a particular skill, craft or profession.

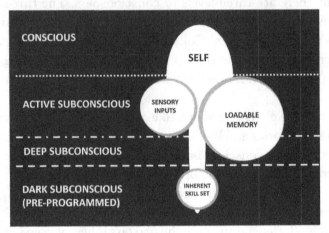

Figure 210

Self.

MORALITY

The morality refers to the decision maker in the brain that references socialization skills, which are generally taught, which include being mindful of one's family and neighbors. Morality represents the essence of knowing right from wrong. See Figure 211. Part of morality is innate and part of morality is taught or learned from social interaction in one's youth. Morality has some deep seeded subconscious roots, such as a parent would be compelled to protect one's offspring from harm. Morality is aided by the fact that in most people following what is considered right or noble or honorable tends to make a person feel satisfied, even if the individual must sustain a sacrifice to carry out the behavior. Morality may be ignored, which can lead to instability, which is often interpreted as guilt. Morality may be superseded by the instinct of survival or by greed. Morality can be modified based on extreme doctrines impressed upon young minds by mentors, or by adverse physical or emotional experiences an individual survived, which molds a person's perception in a negative manner against a person, group, religion, sexual orientation, or country.

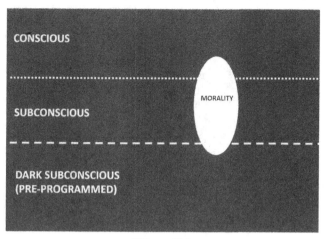

Figure 211

Morality.

A discussion of the definitions and interactions of morality with survival and greed are complex and venture well-beyond the scope of this text. One of the most curious subjects of study and debate is that of how an individual functions when faced with a critical conflict between morality and survival, or morality and greed. Resolution of this matter is left to the storywriters, philosophers and judges. The subject of morality is mentioned in this text only to illustrate that it has its place in the human brain.

SURVIVALIST

The 'survivalist' encompasses the basic qualities needed to survive in a hostile environment. Certain innate algorithms are required to insure proper development and maintenance of the human body. Recognizing food and water, shelter, and acquiring and consuming such items allows for survival. Morality and the survivalist may conflict at times if the drive to survive requires rationing of supplies and overshadows social etiquette. The survivalist may at times also encompasses the sex drive. The sex drive, though generally thought to benefit an individual, actually, probably on the grand scale, favors survival of the species rather than benefit to the individual. The survivalist operates in both the conscious and subconscious portion of the brain, with ties to the dark subconscious. See Figure 212.

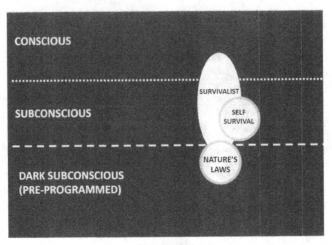

Figure 212

Survivalist.

Much of the behavior allotted to the survivalist portion of the human brain are pre-programmed algorithms. The nature of love, and the means as to how an individual selects a lover has been heatedly debated in romance novels for centuries and is well beyond the scope of this particular text. Romance comes in so many forms it would take volumes of written word to even begin to describe the human experience of love, which in of itself evolves as a person matures through their lifetime. Still, the act of selecting a partner may span behavior as simple as the heated primordial attraction between two people given the right time and opportunity to a much more complex DNA coded attraction between two people. In some circumstances, two individuals may recognize a bonding attraction driven by a dark subconscious coding written into the DNA of their cells. An attraction dictated by an individual's genetic make-up may circumvent manmade laws and social prejudice in favor of Nature's much more powerful primary law of preservation of the human genetic code.

HOTEL MANAGER

The hotel manager portion of the human brain is the pre-programmed human computer which makes human existence possible. The hotel manager portion of the human brain is in constantly optimally coordinating all of the biologic systems required to keep the human body functioning twenty-four hours a day, seven days a week, throughout an individual's lifetime. The hotel manager closely monitors the subconscious systems of the body. The hotel manager informs 'self' that the body is hungry, thirsty, or the body needs to rest, or the body needs to eliminate waste. The hotel manager operates in the subconscious and dark subconscious portions of the brain, sending suggestions to the conscious self as necessary in an effort to optimally maintain proper body function.

The hotel manager (HM) is primarily based in the dark subconscious portion of the human brain. See Figure 213. The HM receives most inputs at the subconscious level. The HM filters the signals input from sensors dispersed all around the interior and exterior of the body, and continuously updates analysis algorithms. Based on preprogrammed algorithms and genetically dictated tolerance limits, as necessary, HM attends to duties around the body in the same subconscious level. The actions of the HM frees the conscious human brain up to function at a higher level of consciousness, rather than constantly having to tend to mundane tasks required to maintain bodily functions at an optimum level. The hotel manager interacts closely with the conscious brain, as necessary, to accomplish a task requiring attention.

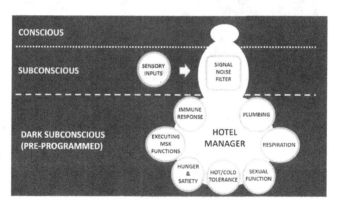

Figure 213

Hotel Manager.

The HM oversees bodily functions such as immune response to infection, executing musculoskeletal functions, assessing and satisfying hunger/satiety, monitoring hot/cold tolerances, sexual function, respiration and plumbing. Plumbing refers to the constant function of the gastrointestinal tract and the urinary tract. For each of the above mentioned subjects, often proper responses dedicated to optimal function requires orchestration of multiple organs located in different sites of the body. It would take volumes of medical books to properly explain how the body manages the various different bodily organs, which is of course beyond the scope of this simplified description.

The scope of function of the HM also includes blood pressure control, blood sugar control, chemical acid/base regulation, thyroid control and the autonomic nervous system. See Figure 214. The autonomic nervous system (ANS) is as the name implies, an automatic nervous system. The ANS is possibly the clearest example supporting the existence of HM. The ANS involves numerous actively working parts including the central (brain) nervous system and peripheral (outside the brain) nervous system interacting with multiple body organs.

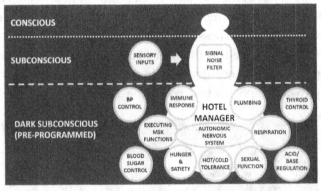

Figure 214

Hotel Manager with extended duties.

The simple act of rising up out of a chair to an erect standing position requires the ANS to orchestrate delicate and highly coordinated function of numerous organs throughout the body. The most noted requirement associated with an individual rising to a standing position is maintenance of an equal blood pressure in the brain throughout the entire event to prevent the individual from becoming dizzy or passing out. Other more obvious functions include the movement of the proper muscle groups in a coordinated fashion in order to accomplish the task of standing erect, increase in heart rate to maintain blood pressure to the brain and peripheral tissues, as well as coordination of vision, cerebellar brain function and vestibular apparatus function in order to maintain balance as the act of standing proceeds.

REGISTERS OF PREFERENCES

There are pre-set values for the human brain's interpretation of various body sensor inputs. These pre-set tolerances are derived from a person's genetic programming. The pre-set tolerances include:

-Range of TASTE tolerances
-Range of SMELL tolerances
-TOUCH/PRESSURE parameters
-HOT/COLD tolerances
-PERSONAL SPACE tolerances
-Range of AUDIO perception, likes/dislikes of sound pattern and intensity
-VISUAL perception, likes/dislikes of visual pattern recognition, types of color
-SAIETY preferences interfacing with PLEASURE thresholds

A few examples:

* Work hard	versus	Tending to play
* Desire nice car	versus	Unenthusiastic about a car
* Gold jewelry	versus	Indifferent about materialistic items
* Desire wealth	versus	Languid about accumulating money
* Careful grooming	versus	Apathetic about appearance

The pre-set tolerances located in the REGISTER of PREFERENCES makes each individual unique compared to every other human. The pre-set tolerances represent a person's likes and dislikes. See Figure 215. The tolerances greatly influence a person's behavior. Tolerances can be modified over the course of a person's life time. Environmental conditions and experiences may alter preferences.

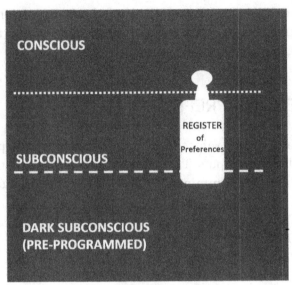

CONSCIOUS

SUBCONSCIOUS

REGISTER
of
Preferences

DARK SUBCONSCIOUS
(PRE-PROGRAMMED)

Figure 215

Register of Preferences

A person may be very attracted by wild flowers. If such a person is stung by a bee while gathering a bouquet of wild flowers, such a negative experience may blunt future desire to possess wild flowers. If the negative experience is significant enough, the desire for wild flowers may be modified from desire to a fear-provoking perception.

If preferences are associated with positive reinforcement, such as the accumulation of wealth leads to increased buying power, which leads to accumulation of more wealth, then positive reinforcement may lead to an addictive behavior associated with a pre-set tolerance. Addictive behavior may not even be clearly evident. An over-zealous desire for alcohol may lead to increased stimulation of the pleasure sensors in the brain, which leads to addictive behavior toward the consumption of alcohol or other drugs.

The pre-set tolerances located in the REGISTER of PREFERNCES are not viewable, except when observing a person's behavior. Pre-set tolerances probably do not follow specific family genetics, but instead are more likely dictated by genetics associated an overall choices of human behavior patterns. Thus, members of a family may have characteristic physical features, but the behavior of the individuals of a family may be quite varied.

The REGISTER of PREFERENCES is illustrated to interact with the conscious brain, as well as the subconscious brain and the dark subconscious. The dark subconscious is where the pre-set tolerance files originate. Within the subconscious, modifications to the tolerances may occur and are stored as active files. The conscious brain may to some extent be able to filter the signals generated by the subconscious and dark subconscious with regards to determining behavior based from these tolerances. Such filtering of signals by the conscious brain may lead to voluntary positive behavioral changes when addictive behaviors are harmful to the individual or persons around them.

CONCIERGE SERVICES: Everyone's Super Power

The concierge refers to the portion of the brain that acts as a gateway to allow for access to special assets of the brain. In a high-class hotel, the person tending the concierge's desk generally is positioned in or near the main lobby of such a hotel. The concierge's desk is generally visible, but the concierge himself/herself generally remains silent unless specifically approached by a hotel guest. The hotel concierge is present to provide information, but usually also has at his/her disposal many powerful assets, which frequently remain hidden, unless a special request is made by the patron to the concierge. Once engaged, an exceptional concierge may be to acquire special opera tickets or schedule a dinner event at an exclusive restaurant or solicit choice tickets to a local sports event or social gathering. The key is knowing the concierge is present and that such a person is willing and capable of assisting hotel guests.

The concierge in the human brain could be said to be similar to a hotel concierge. Only with regards to the human brain, it is a super powerhouse of talent everyone possesses, yet virtually no one consciously realizes such a resource exists. The human brain's concierge exists mostly in the dark subconscious. The human brain's concierge has access to a mathematical processor, access to analytical powers, a random idea generator and to core memory files. See Figure 216. The majority of the concierge services is positioned in the dark subconscious portion of the brain with ties to the subconscious and conscious portions of the brain.

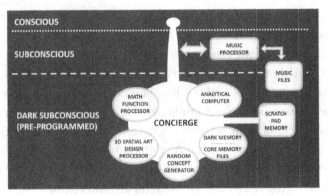

Figure 216

Concierge Services.

SCRATCH PAD MEMORY

Scratch pad memory refers to a form of memory that is a temporary memory unit which stores important memory facts for a limited period of time. Once the utility of the scratch pad is no longer needed, the mental facts held in scratch pad memory are either committed to long term memory or are erased. With respect to the Concierge service this portion of the brain has access to a scratch pad memory unit. A person's conscious state of mind may upload into the Concierge's scratch pad memory details of a problem which needs to be solved. The Concierge's service may add concepts or facts into the scratch pad memory unit in order to compile a solution to the proposed problem. During the problem-solving phase the Concierge service may query the conscious self with ideas in an attempt to fulfill the requirements of the tasked problem.

STIMULATION OF THE CONCIERGE SERVICES

The concierge services may briefly become apparent to an individual at times of crisis or great need, where the conscious portion of the human brain is struggling to accomplish some critical task. The concierge services may provide the conscious portion of the brain a computational analysis of a problem or a mathematical solution to a problem, or simply one or more random concepts which may or may not be helpful in accomplishing the task. In some cases, in times of crisis, the concierge service may function interactively, attempting to provide the conscious brain with a variety of options or solutions to a problem during the timeline the problem confronts the individual. For most individuals the concierge service lays in a dormant state unbeknownst to the individual.

The ability to interact with the concierge service is related to a dedicated effort of learning. The act of studying scholarly subjects, memorizing facts, processing associations and when necessary mastering mathematics equips one to develop an interplay with the concierge service. The functional relationship between self, memory and the concierge's service is dependent upon the amount and diversity of learned information.

People whom recognize the concierge service exists, will knowingly train their brain to actively utilize the processing power of the concierge service. Engineers often interact with the concierge service when problem solving. Often an engineer will load the subconscious brain with a series of parameters associated with a problem, and wait for the subconscious to process the information. The concierge services may respond with a solution in a matter of minutes, hours, days, weeks or even years. Response time is often influenced by urgency to solve the problem, by complexity of the problem and requirement for design of new technology required to solve such a problem. Interaction of the concierge services with an individual may be driven by the level of education of the individual and the experience level of the interval.

Details of the response by the subconscious mind in reference to a solution to a particular problem may be related to whether the solution to the problem can be accomplished by existing technology versus whether the solution requires development of new technology. If existing technology is already available to accomplish the problem, the concierge service may provide very detailed blueprints to solving a problem. If the technology does not exist to solve a problem, the concierge service may provide abstract solutions to a problem. Continued query of the concierge service over time, with trials and failures of existing technology may lead to the concierge service recognizing the existing technology and providing new and innovative ideas to solve a particular problem.

The concierge service exhibits an 'attitude'. As mentioned above, interaction of the concierge services with an individual may be driven by the level of education and experience of the person. The higher the level and type of education influences the technical level at which the concierge service is capable of functioning at with an inquiring mind. The more experience the seeker of knowledge has, the higher the level of productivity the concierge service is capable of functioning at, with such an individual.

Alternatively, emerging brilliance is often stifled and even suppressed by institutionalization. That is, the conscious mind may engage in a learning process at a recognized university and be forced to conform to the current level of understanding of a subject. Conformation is necessary, in order to acquire the basic knowledge of a subject matter, but the need to conform may also constrict inventiveness and creativity. Some learned subject matters fail to advance, simply because generations of student bodies are forced to conform to the will of pre-existing dictums regardless of the validity of dictums. Often the balance between the necessary body of knowledge required to establish and maintain a certain discipline and the coexistence of inventive free thought within such a discipline is delicate, and often collides with catastrophic results inside the embodiment of the hierarchy of an established learning institution.

So, SELF is a matter of combining the elements of appreciation of one's existence and the existence of one's surrounding, with the elements of an inherent skill set, survivalist qualities, morality, a loadable memory, and interaction with the hotel manager services and the concierge services. See Figure 217. Accomplishing self-awareness and functioning as an autonomous individual requires an enormous processing power on the part of the brain. Then to extend the task of self-awareness to facilitate an individual functioning successfully as part of a group of individuals requires even more additional processing power of the brain. To function as part of a social network, an individual's brain must not only analyze and keep track of the needs of one's own body, but also analyze the physical and emotional requirements necessary to optimally function as a productive part of a group of individuals.

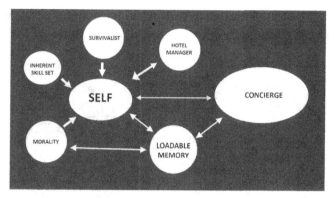

Figure 217

Coordination of services required to produce SELF.

Sigmund Freud described the functions of the human brain in an age that pre-dated computers. In his time, there was no discussion of computer processors, memory files, computer applications, or digital software. In Freud's time, the neuroanatomy of the brain was limited to gross anatomy, that is what could be seen by opening the brain cavity with a surgical scalpel. CT scans, MRI studies, PET scans, and EEG analysis of the brain are all modern inventions. Yet, those having photographic memories obviously store memories as distinct and very detailed files of data, similar to the above-mentioned computer technologies.

GENERAL ANATOMY OF THE BRAIN

The human brain is comprised of multiple complex parts. Presented in Figure 218 is an illustration of the various major centers of the brain which include the frontal lobes, the parietal lobes, occipital lobes, and the temporal lobes. A large fissure runs down the center of the brain dividing the brain into right side of the brain and left side of the brain. The four major lobes of the brain are bilateral, existing both on the right side and left side of the brain. Paradoxically, a person whom is right-handed is generally left brain dominant. A left-handed person is right brain dominant. Men are thought to predominantly use one side of the brain to improve their skill at attending to and successfully accomplishing tasks. Women are thought to more aggressively interact with both sides of the brain as a means of increasing their awareness of their surroundings and to optimize socialization skills.

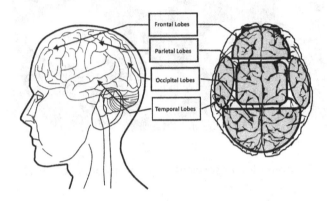

Figure 218

The four major lobes of the human brain including frontal lobes, parietal lobes, occipital lobes, and temporal lobes.

The frontal lobes are located at the front of the brain, right and left sides. The parietal lobes are located top of the head, right and left sides, mid portion of the brain. The temporal lobes are located along the mid portion of the brain, running alongside the parietal lobes. The occipital lobes, are located in the back portion of the head.

Coveted within the central portion of the main eight lobes of the brain are a number of inner core structures. This central portion of the brain includes structures such as the limbic system, thalamus, hippocampus, hypothalamus, and brainstem.[159] The inner core structures are responsible for routing information around the brain. Much of the information is routed from the exterior body back and forth to the brain. Some of the information is routed from the right side of the brain to the left side of the brain. The inner brain components are also likely responsible for the coordinated processing of information, similar to how a central processing unit (CPU) functions within a desktop computer. There is likely a primary Brain Operating Software System (BOSS), which is constructed as a base-four software code, functioning as the bio-computer language platform, responsible for the transfer and storage of information within the human brain. [160] The eight main lobes of the brain, most likely contain individual bio-computer processors, which act in a coordinated effort, possibly at the command of the inner core structures of the brain.

Referred to as the brainstem, comprised of the medulla, pons, and midbrain, is the portion of the brain responsible for the primary muscle motor and sensory nerve signaling to and from the body and head. The twelve cranial nerves allow the human brain to access information regarding the head directly without having to have the signals routed through the spinal cord. Attached to the lower portion of the brainstem is the spinal cord.

The spinal cord is comprised of tracts of nerves flowing down from the brain centers through the brainstem to the scalp, neck, arms, chest, abdomen, pelvis and lower extremities. The nerves of the spinal cord are responsible for sensory information being routed up from the body to the brain, and for instructions being sent from the brain to the peripheral components of the body. A variety of sensors located on the skin and within vital organs, keep the brain constantly apprised of the condition of the body. Instructions from the brain drive muscle function, control blood pressure, routing of blood flow, lung function, renal function and many other physiologic processes to keep the body operating at optimal proficiency.

The customary eight lobes of the human brain offer various sites of specific mental function. The frontal lobes are thought to harbor the skills of consciousness, awareness of self, creativity, artistic skill and emotion. Between the frontal lobes and the parietal lobes, on each side, is located Broca's area. This portion of the brain is responsible for the production of speech.

The parietal lobes are anatomically divided front and back by a noticeable fissure, which produces a physical gap between the two parts. The back portion of the parietal lobes generally is the part of the brain that receives sensory inputs from the body's extremities. Sensory inputs include fast/slow pain, pressure, vibration, hot/cold sense, and proprioception. Proprioception is the sense of knowing where in three-dimensional space an individual body part is in relation to the rest of the body. The front portion of the parietal lobes is where movement of the head and extremities is designed, calculated and coordinated. This area is referred to as the cortical homunculus. Movement of any portion of the body generally requires multiple muscles to work together either in succession or in combination, in real-time, which often requires numerous signals to be transmitted down the spinal cord to the muscles to carry out the desired movement of the body part. The task becomes even more complicated regarding coordinating the sensory inputs and neurologic output signals when several, if not all, of the body parts are required to execute a particularly complex task, such as an individual performing a cartwheel or a back flip.

The temporal lobes are generally thought to harbor the qualities of auditory interpretation, mathematical skills, and language development. A portion of the brain known as Wernicke's area, is located in the back portion, also referred to as the posterior of the temporal lobe and parietal lobe. Wernicke's area is responsible for constructing language for purposes of verbal communications. Once words have been strung into a single expression or into sentences, this information is forwarded to Broca's area. Within Broca's area instructions for the physical formulation of words are generated. From Broca's area, these muscle commands are sent to the brain stem, then down the spinal cord to stimulate the muscles of the neck and lungs, which control air crossing over the vocal cords, in order to carry out the process of speech.

The occipital lobes, located in the back of the human brain, receive data inputs from the eyes, by means of the optic nerves which stretch from the back of the eyeballs to the rear brain compartment. The occipital lobes coordinate and interpret images reflecting radiant energy in the visible light spectrum. An individual interprets what the eyes see, utilizing the processing power of the two occipital lobes.

Underneath the occipital lobes is the cerebellum. The cerebellum is housed in a cavity of bone, deep in the lower, back portion of the cranium. The cerebellum assists in coordinating overall body balance and muscle movement.

NETWORK OF BIO-COMPUTER PROCESSORS

It is likely each major portion of the human brain houses at least one independent processing center. Such bio-computer processing areas are also likely to have dedicated long-term or short-term memory storage units. The frontal lobes likely contain several bio-computer processors. See Figure 219(a). The forward and back portion of each parietal lobe likely has dedicated processors. Each temporal lobe likely has several processors, including one for auditory interpretation, one for mathematic skills, and one for Wernicke's area. Broca's area likely has a dedicated processor. The occipital lobes likely have graphics processing capabilities. The cerebellum is likely a three-dimensional processing in order to coordinate balance. This central or inner core portion of the brain including the limbic system, thalamus, hippocampus, hypothalamus, and brainstem are likely a combination of central processing capability versus information routing units. See Figure 219(c).

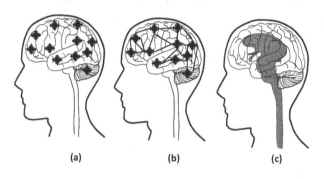

(a)   (b)   (c)

Figure 219

(a) Biocomputer processors likely are positioned in critical areas of the human brain. (b) Likely an individual's talents are related to how the processors are hardwired together. (c) The internal core of the brain, the midbrain and brainstem likely house the central processing unit of the human brain.

The various processing centers of the human brain are intricately connected. The 'individuality' of human brain function is likely related to how the individual processors are hardwired together and how the individual processors are hardwired to the internal core of the brain. See Figure 219(b). The numerous individual bio-computer processors comprising the human brain are connected by nerves to each other in a complex computer network. How large the nerves are connecting various processing centers, probably dictate the natural talents individual people demonstrate on aptitude tests. Depending upon how the nerve connections are constructed in the linking of processing centers, likely dictates those individuals whom are proficient in operations such as mathematics, mechanics, artistic crafts, memorization, articulation of speech, visual recognition, athletic talents.

AREAS OF DEDICATED BRAIN FUNCTION

The most exterior portion of the human brain, which is also associated with higher intelligence in all humans, is divided into the specific regions. Given that there are a number of processors scattered around the brain, these processors have specific functions. The various areas of the brain have been determined to harbor various processing centers to include cranial nerve inputs, peripheral nerve inputs and outputs, judgement, emotions, creativity, speech centers, auditory center, math skills, balance and sight centers, 3D spatial awareness. See Figure 220. There are numerous other processing centers throughout the human brain, this list and associated figure identifies only a few regions of the brain.

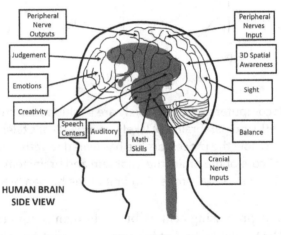

Figure 220

The human brain is comprised of numerous processing centers with dedicated services associated with different areas of the brain.

The various processing centers comprising the human brain allow humans to work on a number of mental tasks simultaneously. Dedicated processing centers also allow for coordination of efforts to complete tasks successfully. It is likely the mid brain, seen as a gray silhouette in Figure 219, acts as the central processing unit, to coordinate the activities of the various processing centers.

To leap across a creek while hiking in the woods, jumping from one side of the creek to the other, requires the brain processing for vision, balance, muscle coordination, cardio-pulmonary support; which requires the visual center, the cerebellum, an orchestration of numerous muscle commands, and the heart and lungs to all be coordinated in the effort simultaneously. If, one's brain is not successful in coordinating the processing efforts of the various individual brain centers, then one lands in the creek and ends up with their shoes and socks soaked for the remainder of the hike.

Hardwiring of the brain with nerve connections is dictated by the construct data contained in an individual's genetics. The learning process of having a child, teen and young adult attend school, uploads information into various portions of the brain. Language is stored in Wernicke's area. Visual recognition data is stored in the occipital lobes. Mathematic skills, such as equations and constants and their applications, are stored in the temporal lobes. Art techniques are stored in the frontal lobes. Three-dimensional design and problem solving are conducted in the region between the parietal lobes, posterior temporal lobes and the occipital lobes. The learning process not only uploads data into memory areas, but depending upon the type of schooling, also opens or reinforces nerve connections between specific processing centers. Students of business, science, the arts, social studies and engineering utilize differing portions of the human brain in order to develop and optimize their skill sets in order to become proficient in their particular discipline. Where scholarly students develop conscious and subconscious neuropathways in their brains tailored to the needs of their professional work, professional athletes work with the parietal lobes and perfect the brain commands and muscle memory associated with the muscle groups required to participate in their chosen sport.

REVISITED: SELF, MORALITY, SURVIVALIST, CONCIERGE, HOTEL MANAGER

Given the human brain is comprised of a network of processors, coordination of the use of the varied biocomputer processors leads to overall function of the brain. SELF is likely established by the function of the processors located in the frontal lobes of the brain. MORALITY is likely related to coordination of the processors in the frontal lobes and the limbic system. SURVIVALIST's problem-solving skills are likely related to coordinated processing of the temporal, posterior parietal and occipital lobes. CONCIERGE is likely deep rooted in the inner core of the brain with access to restricted dark subconscious memory files. The HOTEL MANAGER is likely a complex coordinated effort of numerous processors and information routers located in the central brain, which function autonomously to optimize body function and management of resources, so that the conscious thought is not burdened by such maintenance requirements, unless crisis occurs, at which point the conscious mind is alerted to the problem.

MODES OF THE BRAIN-BODY

The brain and body may participate in various modes. While in a particular mode the brain and body act in a highly coordinated fashion to accomplish a task or series of tasks. Essential functions of the body are maintained, in some situations optimized, where nonessential functions are minimized or put on standby. Such modes include:

COURTSHIP OF A MATE
PARTICIPATING IN SEXUAL FUNCTION
ATHLETIC TRAINING/COMPETITION
FIGHT OR FLIGHT RESPONSE TO A THREAT
PERFORMANCE OF A WORK-RELATED DUTY
CONSUMPTION OF NUTRITION
STUDY
PURSUING A HOBBY/PARTICIPATING IN A GAME
PRESENTATION TO AN AUDIENCE
OPERATION OF A VEHICLE
EXPLORATION OF A NEW ENVIRONMENT

A mode is a unified function of the assets of the brain combined with necessary body parts to accomplish a goal or complete a task, the result of which has significant value to the individual. The brain suspends nonimportant thought processes (such as daydreaming) and dedicates analytical resources and devises muscle commands to enact the proper muscle groups to perform required acts. During a mode, the brain and body are dedicated to the task at hand, acting in as close to unison as possible.

VARIED SCHOLARLY USES OF THE HUMAN BRAIN

Thought not necessarily obvious, the manner which differing individuals utilize their brains may be astoundingly varied. The thought structuring utilized by a physician versus an engineer are nearly polar opposites. Physicians approach a problem in a significantly different manner than an engineer would approach the same problem. Obviously, the fact base knowledge learned by a physician during medical training is quite different than the mathematic-based curriculum taught to engineering students. Upon completing their respective training, both disciplines would inherently have differing skill sets with which to solve problems. But even if the baseline knowledge were equalized, still the approach to solving a particular problem is quite different utilized by a physician versus the method utilized by an engineer to solve the same problem.

THE PHYSICIAN

The approach to learning utilized by physicians tends to incorporate the front lobes of the human brain. See Figure 221. Physicians tend to place in their memory an enormous amount of medical facts. The medical information a physician learns starts with human anatomy and proceeds to include chemistry, molecular biology, information regarding disease states, histology, pathophysiology, pharmacology and surgery.

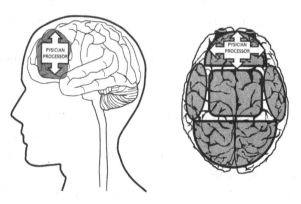

Figure 221

Physician's brain predominantly utilizes frontal portion of the brain.

At the beginning of medical training, a medical student tends to concentrate on the study of volumes of facts regarding human anatomy and human health. As a student doctor proceeds in his/her training, like facts are arranged into memory files. As a physician transitions from a student to clinical practice, the fact base knowledge is further refined and arranged into numerous decision trees. See Figure 222. In order for a student to evolve into a clinician and be able to translate volumes of facts from what has been learned into the skill of recognizing disease processes in afflicted patients and determining therapeutic interventions elaborate decision trees are required to be memorized.

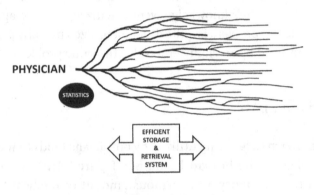

Figure 222

Physician's brain utilizes decision trees.

During the study of medical information and the execution of the learning process required to memorize volumes of medical facts, students whom become physicians develop a very efficient information storage and retrieval system in their brain. Medical students undergo study of vast quantities of medical information and are generally required to frequently demonstrate that they have properly memorized such information by successfully passing knowledge-based tests.

Physicians rely heavily on the self-portion of the human brain and the loadable memory attached to the self-portion of the brain. See Figure 223. The loadable memory attached to the self-portion of the human brain is at least capable of retaining a majority of the information required to become a clinician. The decision trees learned in the art of medicine are seemingly endless in the information which can be attached to a particular branch of the decision tree, as well as new branches seemingly can be added at any time without having to delete previously stored decision tree branches.

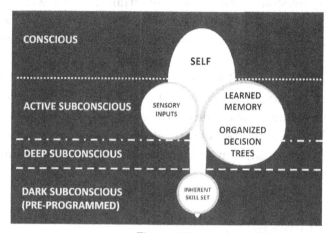

Figure 223

SELF includes loadable memory, sensory inputs and an inherent skill set.

Generally, to be a physician, the inherent skill set attached to the SELF portion of the brain needs to be able to process and retain large quantities of information. Medical schools tend to therefore use performance in undergraduate classes requiring similar efficient storage and retrieval skills, including study of science and foreign language, as a means of predicting success in the learning process required of a medical student.

Physicians tend to function in an environment where what they encounter, in clinical practice, is most often something they have seen before in their medical studies. A majority of the time the physician bases decisions regarding diagnosis, treatment strategies and pharmaceutical intervention on the decision trees they have previously learned in a like manner to other physicians across the educational process. This memorization process of like medical knowledge is often referred to as the 'standard of care' approach to medicine, where all like medical problems are to be treated similarly. Patients are referred to specialists when needed, due to a specialist having received additional training or experience, which results in a specialist having acquired a more elaborate set of decision trees pertaining to a particular subject matter compared to that of the general clinician.

Often physicians are required to memorize volumes of interrelated facts, though how the facts are tied to together is sometime unknown or poorly understood. Reasoning is important to a physician, but often though, sophisticated medical science often lacks the means of explaining how certain medical phenomena fit together in a disease process. An example would be sarcoidosis. A patient may present with bumps under the skin and swelling involving the ankles, and the physician is required to process this information to arrive at the possibility of swelling involving lymph nodes in the chest, which is highly suggestive of sarcoidosis. The etiology of sarcoidosis and how this disease process may involve the chest, skin and joints at the same time remains elusive to those practicing clinical medicine.

THE ENGINEER

Engineers generally work in an environment where they are often invoking logic and solving problems no one has ever seen or encountered previously. Engineers routinely are required to take learned skills and adapt them to new conditions. The engineering curriculum tends to heavily utilize mathematical modeling. Engineering students are generally taught a variety of problem-solving methods and mathematical modeling algorithms. The memorization of volumes of facts does not work as well to create engineers as is works to generate physicians. When no one has ever encountered a problem an engineer may be tasked to solve, the memorization of facts alone does not equip the individual to derive ingenious solutions to tough problems.

Mathematical modeling provides a means of predicting outcomes given a set of parameters, to allow an engineer to arrive at optimal solutions to a problem prior to committing valuable resources in the construction of projects. Far better to utilize mathematical modeling to predict the maximum loading a bridge can sustain prior to construction of a bridge, then construct a bridge and have it collapse due to the weight of cars and trucks that might utilize the bridge.

In addition to acquiring a variety of sophisticated math skills, engineering students often are subjected to having to think and work in three-dimensional space. The recent progression of computer technology to make available computer assisted modeling (CAM) software, where three-dimensional imagery of objects can be studied and manipulated in real time, has dampened the need for engineering students to think in three-dimensional space on their own. Still, the engineering student whom is capable of developing the skill of thinking in three-dimensional space within the context of their brain possess an enhanced skill of conceiving concepts prior to committing resources to a project.

To accomplish thinking in terms of three dimensions, engineers utilize a different portion of the brain than physicians. Where physicians utilize the front portion of the brain, engineers utilize the back portion of the brain. The mid portion of the brain is called the parietal lobes. The back portion of the parietal lobes is where the sensory inputs from around the body are congregated. Behind the parietal lobe is the occipital lobe. The occipital lobe is where the human brain processes vision.

Vision is accomplished by the eyeballs at the front of the head absorbing light and sending electrical impulses by means of the optic nerve to the back of the head to be interpreted. Eyeballs are approximately elliptical fluid filled spheres. The front of each eye is positioned a lens. The lens focuses light toward the back of the eyeball. At the rear of the eyeball is positioned the retina. Light taken in by the lens of the eyes traverses the interior fluid of the eye. The cells in the retina absorb the light input and transforms the light signals into electrical impulses. Cones are cells that absorb light during the day and are able to distinguish color. Rods are a type of specialized cell used for night vision and generally absorb and detect shades of gray. The electrical impulses generated by the retina are transferred across the right and left optic nerves to the back of the head to the two occipital lobes. It is the function of the right and left occipital lobes to interpret the light signals the eyes detect. Objects the eyes gather light from and the occipital lobes detect are given names and meaning as a part of the learning process throughout an individual's lifetime. The brain cells interpret light signals as vision.

The temporal lobes are finger like structures located on the right and left side of the brain at about the level of the ears. There are twelve cranial nerves. The eighth cranial nerve is responsible for hearing and transfers signals from the ear on each side of the head to the temporal lobe. The temporal lobe of the human brain is thought to be responsible for interpreting sounds including music, and to act as the processing center for mathematics.

The cerebellar portion of the brain is located in the back of the head, underneath the primary lobes of the brain. The cerebellar portion of the brain is responsible for balance and position sense. A bony cavity protects the cerebellum.

The three-dimensional thought processing an engineer engages in is located in the region of the human brain which is boxed in between the posterior parietal lobes, the occipital lobes, the temporal lobes, and just above the cerebellum. The three-dimensional imaging in an engineer's brain is accomplished by coordinating the sensory inputs which stimulate the parietal lobes, with the imagery stored by the occipital lobes, with the mathematical skills conducted in the temporal lobes and with the sense of balance conveyed by the cerebellum. See Figure 224. An engineer is able to view the image of an object inside his or her head in a spatial design center, as well as change the shape, alter perspective, and change color of the object. At times the engineering brain might deconstruct an object and rearrange the pieces, deleting some pieces, and adding different pieces in an attempt to optimize a new design. Seeing objects inside their heads in three-dimensions allows the engineer to solve problems requiring three-dimensional solutions.

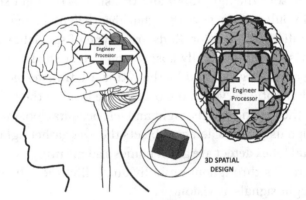

Figure 224

Three-dimensional spatial design processing utilized by an engineer.

A truly skilled engineer adds a fourth dimension to their analysis, the dimension of time. An experienced engineer may be able to take a three-dimensional object that can be seen inside their head and run it through a time course in an attempt to predict what may happen to the object given a variety of parameters or test exposures to create predictive models. Engineers would then take the approximations generated by their brain imagery and produce physical models, mathematical models and/or computer models to test out the theories posed by their brain's processing power. Some parties would possibly refer to an engineer's skill of generating new ideas in the three-dimensional spatial design area of the brain as simply 'imagination'.

Likely anyone can take advantage of the three-dimensional processing inside their brain. Engineers tend to be forced to train their conscious mind to operate in this region of their brain by nature of the requirements of an engineer's training curriculum and later their professional duties. It would be assumed that craftsmen and tradesmen of all kinds, individuals whom use their hands to conduct work, would utilize similar brain functions to operate machinery and conduct the duties of construction work. Artists and sculptures would likely also utilize the three-dimensional region of their brain to envision their work as they craft their artwork.

Thus, a physician's brain functions in a vastly different manner than that of an engineer. Physicians memorize volumes of medically related information and formulate elaborate decisions trees from which they diagnose disease states and initiate medical therapy to manage the disease states that are detected. Instead of memorizing vast amounts of information, engineers memorize skills which assist in analysis including geometry, trigonometry, calculus, vector analysis, programming algorithms and Laplace and Fourier transforms to name a few. See Figure 225. Specialized engineers tend to learn and become most acquainted with analysis skills tailored to their form of work.

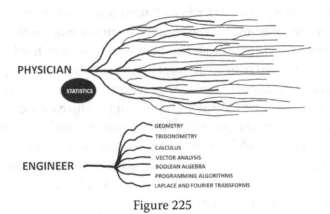

PHYSICIAN

STATISTICS

ENGINEER

GEOMETRY
TRIGONOMETRY
CALCULUS
VECTOR ANALYSIS
BOOLEAN ALGEBRA
PROGRAMMING ALGORITHMS
LAPLACE AND FOURIER TRANSFORMS

Figure 225

Differing organization patterns between physicians and engineers.

Surgeons are physicians whom primarily conduct surgery. The human body is a three-dimensional entity; therefore, a surgeon is faced with three-dimensional problems when it involves surgery. A surgeon's brain functions as a combination of a medical physician and engineer. Surgeons generally memorize human anatomy, memorize disease processes similar to the medical physician, and memorize and follow the procedural steps necessary accomplish a given surgical task. Given the three-dimensional aspects of the human anatomy, the surgeon also needs to invoke and actively utilize a conscious three-dimensional imagery and processing located in the posterior portion of the brain.

THE PHYSICIST

Physicists represent a high order of mathematical science. Where engineers often rely on conceptualization of existing objects, physicists work in the realm of theoretical entities such as isotopes of atoms, as well as subatomic entities such as protons, neutrons, electrons, and other particles. Given only the tracks of subatomic particles have been seen and not the actual physical entities themselves, a physicist is at the disadvantage of studying the characteristics of particles which are frequently moving at the speed of light. A physicist's brain is heavily rooted in mathematical analysis and mathematical modeling of the behavior of these fleeting building blocks of the universe. See Figure 226.

Presumably the physicist would be extensively utilizing the temporal lobes of the brain, where the math centers are located.

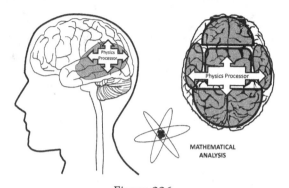

Figure 226

Illustration of concept brain of a physicist.

THE ARTIST

A person with talent for the arts, harbors a brain function which is a blending of front and back portion of the brain. See Figure 227. An artist intertwines emotional feelings processed in the front portion of the brain and limbic system, along with the three-dimensional spatial processing afforded by the brain's graphic design center nestled in between the posterior parietal lobes, occipital lobes and posterior temporal lobes. Artists are able to visualize their creative work in the graphic design center, whether it be drawing, painting, sculpturing, or music composition, and temper their work based on a set of emotions or experiences the artist wishes to express. Once a creative work has been perfected in the graphics center, the frontal lobes and anterior parietal lobes work in unison to create the muscle function to make the art form materialize in whatever medium the artist views the work to provide the optimum impression on the intended audience.

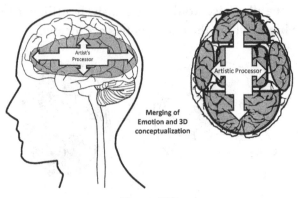

Figure 227

Illustration of concept brain of an artist's brain.

CORE MEMORY FILES: Echoes of the Ancient Designers

Core memory files, are likely to be the most controversial of the subjects discussed in this entire text. Core memory files are located in the dark subconscious and accessible only through querying the Concierge services of the human brain. See Figure 228. Core memory files are accessed in relation to training, to need to know and in many cases related to music. SELF portion of the brain, can in unique circumstances, access a limited portion of the core memory files.

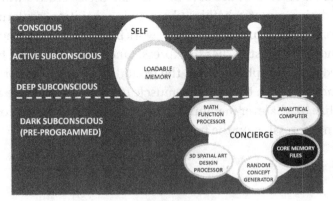

Figure 228

Concierge services has access to core memory files.

Core memory files refers to pre-programmed information stored in the deep recesses of the human subconscious. These dark files are echoes of the ancient civilization which designed Mankind and the ecosystem which has flourished on earth over the last 4.3 billion years. Pre-programmed information would include detailed designs of the technology mankind has developed over the ages. One might guess the blueprints of a galactic starship might even be present in the average human brain. Access to such information is limited, but generally attainable, as mentioned above in relation to (1) type and level of training, (2) a need to know and (3) the level of sophistication of music at the time the query is made. If an intelligent lifeform were to have created what we recognize as organic life, then more than likely, such an advanced intelligent life gave humanity intellectual tools with which to work with to be utilized to create technology to solve problems.

Most would be skeptical of the existence of genius memory files. This is rooted in the fact that 'evolution' has generally been taught as the science-based principle behind the driving force to explain the existence of life on earth. The story of evolution generally starts with a pool of water teaming with the building blocks of organic molecules, which is then struck by lightning causing molecules to arrange themselves into structured forms. Then, given enough time, generally referred to as billions of years, various combinations of molecules randomly interact, creating the building blocks of life both DNA and RNA. Again, provided enough time, the building blocks of life both DNA and RNA randomly combine themselves into highly complex chains forming the blueprints of life, resulting in the construct of living organisms. Evolutionists never venture to explain where the abundance of water covering the surface of the earth came from at the time life began, nor how earth's atmosphere happened to become dominated by nitrogen, when the atmosphere of all of the other planets in the solar system is predominantly comprised of hydrogen and helium. Those whom teach evolution also tend to fail to explain where the machinery necessary to read the DNA (polymerase I, II and III molecules) and the machinery required to edit the precursor RNA molecules (spliceosomes) come from. Oh well, another sleight of hand magic trick.

To subscribe to the theory of evolution means that one must accept that the entire complexity of life and the elaborate ecosystems which fill the earth, came to be, by no means of logic or planning, but by shear random luck. Much of the content of the text is rooted in physics. To a student of physics or engineering, it would seem difficult to accept that such a complex form as the human body with all of its interactive systems and complex feedback loops, would be the result of random events or that the life on earth, in all of its varying forms, was created at the whim of some deity. Logic would dictate that life is due to an organized preprogrammed biologic computing root cause, which is still actively at work today, analyzing environmental factors and modifying life to survive given the ever-fluctuating dynamics of the environment.

Pushing the envelope further, the presence of pre-programmed intellectual tools provided by the designers of life, to assist in development of technologies, maybe difficult to accept. Still, we see such intellectual files at work every day without doubting the existence of such files. Such examples of pre-programming intelligence can be seen in birds building elaborate nests, beavers constructing dams, spiders spinning elaborate webs, squirrels foraging for nuts to save for the winter, animals and birds of all kinds breeding and caring for their young. Innate animal intelligence, is a subject we generally do not venture to explain, but instead simply refer to and blindly accept as 'animal instinct'.

The process of a caterpillar transforming into a butterfly defies the theory of evolution. Metamorphosis, where one form of life turns into a completely different form of life, sets in motion the concept that it is illogical to think that random circumstances, errors in the DNA no matter how many times such mistakes might occur, led to such a highly detailed process. It is hard to conceive that 'random errors' led to a ground-based caterpillar transforming into a butterfly, a creature with wings with the capability to fly. Metamorphosis is arguably a clear example of a designed pre-programmed process written into Nature's DNA. The magnificent transformation of a caterpillar to a butterfly, may be one of the most obvious clues instituted by the original designers of life, to inspire us to seek out the details of the DNA's programming.

A number of notable humans have had visions of technology which were based on mechanical principles well beyond the science of their time. For those who would doubt the existence of core memory files, one only has to study the life of Leonardo DaVinci, the great sculpture and artist. DaVinci lived 1452-1519. Most noted for his art talents, he also envisioned and drew detailed drawings of flying machines, submersibles, artillery pieces, and even a mechanical tank, centuries before any of such machines could be properly constructed.[161,162] Dmitri Mendeleev (1834-1907), a Russian chemist and inventor, formulated the periodic law and created a version of the periodic table was quoted as saying 'I saw in a dream a table where all of elements fell into place as required.'[163] Many prominent figures throughout the ages have had access to core memory files, and have exploited the information stored in such files to advance the human technological presence.[164]

Still, one could describe conceptualization of innovative technically important concepts as intuition or intelligence or again, random luck. An individual dreaming of interstellar space travel, years before such travel is possible, is not necessarily representative of access to core memory files. Therefore, there is little proof that any pre-programming has existed, much less contributed to the origin or development of life on earth...until you study the HIV genome.

HIV GENOME CODING PHENOMENON

Human Immunodeficiency Virus (HIV) is the result of a string of 9716 nucleotides which comprises the virus's genome. Within this segment of genetic material lie all of the blueprints to construct the virus, which includes instructions to make the RNA, the enzymes which convert RNA to DNA, enzymes to insert the viral DNA into the human genome, construct of the outer shell of the virus, and the receptors which mount on the surface of the virus's outer shell so that the virus can locate a host cell within which to replicate. If, the HIV genome was decoded by a linear read method (decoding all of the nucleotides sequentially), then one could argue that the 9716 nucleotides, which comprise HIV's genetic blueprints, were somehow randomly organized by fate. But HIV's genome is not decoded linearly. HIV's genome is far more sophisticated and abstract.

Human derived computer technology has been based on binary computer hardware and software. The root of all computers is a base-two language of ones and zeros. The HIV genome is a clear example of utilizing both a base-three language and a base-four language in tandem.

Within the HIV genome there are nine segments of nucleotides where a frame-shifting is used in the base-three code to facilitate the same segment being read at least twice, resulting in two differing decodings of the information. HIV represents a data compression technique utilizing a combination base-three and base-four languages that is far superior to any electronic computer data compression technology in existence today. HIV was not contrived as the result of random occurrences. Some form of intelligence, far greater than current human intelligence, constructed the HIV genome. Detailed analysis of the HIV genome was presented at European Human Genetics Conference 2017, Copenhagen, Denmark. See Figure 229.

Figure 229

Presentation of detailed analysis of HIV genome demonstrating a highly complex preprogrammed base-three/base-four data compression technique at nine locations within the virus's DNA.

ESHG Abstract 2017, Copenhagen, Denmark

Optimizing data compression technique determined by decoding HIV-1 HXB2 DNA genome, assists in deciphering the means the human genome is transcribed and translated

An optimizing data compression technique involving multiple genes is revealed in analyzing HIV-1 HXB2 genome K03455.1. Understanding HIV's data compression assists in an expanded understanding of the human genome. The data compression technique ingeniously utilizes features of the codon code and nucleotide code to generate overlapping genetic information, which facilitates 8 segments of HIV's pre-mRNA code to be read twice producing differing mRNA sequences each time the pre-mRNA code is deciphered. One segment of viral genetic sequencing is read three times, interpreted differently each time it is decoded.

The HIV-1 HXB2 genome k03455.1, 9719 nucleotides, comprised of 9 genes including gag, pol, vif, vpr, tat, rev, vpu, env, and nef, which produce 9 master proteins designated as gag PRO_0000261216, gag-pol PRO_0000223620, vif PRO_0000042759, vpr PRO_0000085451, tat PRO_0000085364, rev PRO_0000085279, vpu PRO_0000085433, env PRO_0000239240, and nef PRO_0000038365.

Demonstrated in Table 1, is evidence of a frame shift involving each gene. There are 8 segments where codon code overlaps, which facilitates the HIV genome to be 825 nucleotides shorter than if such a data compression technique were not utilized. Env, tat-2, and rev-2 all share a 45 nucleotide sequence. This data compression technique represents a higher order complexity in the construct of the genome, far more sophisticated than a simple linear read, which facilitates the HIV genome to precisely fit inside the HIV virion.

Since HIV genome utilizes the same genetic cellular machinery as human genes, recognizing the microcosm of data compression technique present in the HIV genome broadens the analytical approach to deciphering how the spliceosome and ribosome complexes interact with human DNA, and will stimulate innovative human genetic therapies.

Gene 1 Protein 1 Gene 2 Protein 2	Start-Stop Positions	Point of Frame Shift # Nucleotides that Overlap	Nucleotide Code at Point of Frame Shift	Codon Code at point of frame shift
gag 261216 gag-pol 223620	790 - 2292 2091 - 5096	2091 202	aat ttt tta ggg aag... ...agg gaa gat I 2091	NFLGK.... ...RED... ^ 2091

gag-pol 223620 vif 042759	2091 - 5096 5041 - 5619	5041 / 56 5041	gat tat gga aaa cag... ......atg gaa aac aga I 5041	DYGKQ..... ....M(start)ENR ^ 5041
vif 042759 vpr 085451	5041 - 5619 5559 - 5850	5559 / 61 5559	gat aga tgg aac aag.... ...atg gaa caa gcc I 5559	DRWNK ....M(start)EQA ^ 5559
vpr 085451 tat-1 085364	5559 - 5850 5831 - 6049	5831 / 20 5831	gca aga aat gga gcc... .....atg gag cca gta I 5831	ARNGA... ....M(start)EPV ^ 5831
tat-1 085364 rev-1 085279	5831 - 6049 5970 - 6045	5970 / 80 5970	ggc atc tcc tat ggc agg... .....atg gca gga I 5970	GISYGR ....M(start)AG ^ 5970
vpu 085433 env 239240 *33-856	6062 - 6310 6225 - 8795	6062 / 86 6062	agt ggc aat gag agt... .....atg aga gtg I 6062	SGNES ....M(start)RV ^ 6062
env 239240 rev-2 085279	6225 - 8795 8379 - 8653	8379 / 275 8379 *combine 'a' from rev-1, 'ac' from rev-2	tcg ttt cag acc cac... .....((a)ac) cca cct* I 8379	SFQTH ...(N)PP* ^ 8379 *combined 'aac' codon for 'N'
env 239240 tat-2 085364	6225 - 8795 8380 - 8424	8380 / 45 8380	ttt cag acc cac ctc... .....ccc acc tcc I 8380	FQTHL PTS ^ 8380
env 239240 nef 038365	6225 - 8795 8797 - 9417	8797 / 0 8797	att ttg cta taa... ...atg ggt ggc I 8797	ILL(stop).... ...MGG ^ 8797

Table 4

Overlapping genetic code, utilizing a base-3 code inside a base-four code, present in the HIV genome, argues that the programming features of this viral genome were most certainly designed.

MUSIC, THE VARIABLE CONSTANT

Music is an example of a variable constant, and is the variable constant which is critical to the technologic development of the human race. See Figure 230. Many works of music contain musical keys, and these keys when heard by an individual with a prepared mind, will open access to portions of the brain that will introduce new ideas to the individual. Louis Pasteur is quoted as stating 'Chance favors only the prepared mind'. [165] This is a true statement in degrees. If an individual has prepared their mind through studies and/or experiences, they are more likely to take advantage of the opportunities afforded them, compared to an individual whom has not invested so vigorously in the process of learning or possess lesser experience and thus is not able to equally comprehend new ideas or interpret old ideas given new challenges.

Figure 230

Music is the variable constant which unlocks core memory files.

Music changes as technology changes. As the technology becomes more sophisticated, the musical instruments change and become more sophisticated. New developments in music do not follow a strict timeline for advancement, but instead follow the progress of a 'technology' timeline and artistic skill, which facilitates freedom of expression of music, changing music as technology changes. For some, the human ear may detect patterns in music, and in some, may unlock core memory files in a sequential order. Those whom have the academic training may optimally be able to access and utilize such core memory files. Again, the concierge services, which controls access to core memory files, maintains a low profile and responds to the conscious SELF of an individual's brain only when specifically queried on a particular matter and then releases only limited files as required.

Certain compositions of music contain musical keys. A musical key would be a signal embedded in the construct of the music which acts as a subconscious signal to the brain. Such musical key signals may unlock dark subconscious memory files.

A music piece containing a musical key is comprised of three parts. The first part is an introduction that instructs the human brain to pause and listen. The music tells the brain that the following message is not random noise, but instead an organized signal and that the brain should prepare itself to receive the musical key. The second part is the musical key. The third segment of the musical piece tells the brain the musical key has ended and to return to activities of daily life.

ADVANCED CREATIVITY

The process of learning creates 'applications subroutines' (Aps) in the brain that can act as tools the brain is able to use. These subroutines are stored in the subconscious. The more Aps the brain has to utilize, the more powerful and versatile a person's brain becomes regarding its ability to solve problems. The basis of higher learning is language. The next layer is figure associations. Math is more complex. The sciences offer depth of knowledge. The study of music and math assist in learning to organize information in the brain. Memorization, as painful as it sometimes is to learn a particular subject matter, offers means to quickly access important data that requires rapid retrieval. Complex analysis techniques are learned with a blending of learned knowledge and experience in a particular field. For those who master two opposing fields of study, this assists in the person blending the subjects in novel ways. Participating in a combined learning process provides varied perspectives to subjects which increases the opportunities to make advances in one or both fields.

So, in a scholarly individual, access to core memory files can be enhanced to solve problems. The acquisition of detailed learning sets the foundation or understanding the information released by the core memory files. Once a person recognizes core memory files exist, a person can ask the subconscious for information regarding a particular subject. The subconscious brain may first analyze whether the individual has a legitimate reason to access such information. If the requestor is legitimate, the subconscious brain may query the music files to see if the timing of the release of such information is properly aligned with the current state of technology. New technology released too soon, when materials and methods are not available to construct the new technology is considered 'science fiction' and generally is not associated with proper advancement of technology. New technological concepts released when materials and constructions methods make the technology practical offers optimal use of resources and advancement of Mankind.

If a scholarly person has a legitimate reason to access new concepts, has acquired the proper training to utilize the technology, and the state of music suggests the technology can be successfully utilized, the concierge service may release to the conscious SELF portions of memory files from the core memory files. See Figure 231.

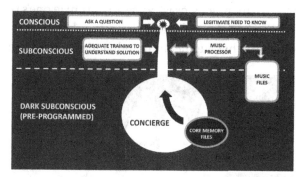

Figure 231

Core memory files released when proper requirements of legitimate need to know, adequate education and proper advancement of music have all been satisfied.

Some individuals have either trained their brain or innately have a brain hardwired to have exceptional access to the core memory files. Such individuals are often referred to as a genius. It would be assumed that anyone might be able to access the core memory files inside their head given the proper training and desire to access such files. The fundamental means to access such files is primarily the knowledge that such pre-programmed memory files exist in everyone's brain.

IN SUMMARY

To recap, the brain supports differing levels of consciousness to include the conscious, subconscious and dark subconscious. There are a number of actions the human brain participates in to include Self, Survivalist, Morality, Hotel Manager, Concierge Services, Register of Preferences. See Figure 232. The human brain is comprised of a complex network of bio-computer processors. Each processor serves a differing role in function of the body and academic aptitude of the individual. A Brain Operating Software System is utilized to capture sensor data from around the body, transfer information throughout the peripheral and central nervous systems, and store pertinent information in short-term or long-term memory. Network design of the hardwiring of the brain's nervous system tends to dictate a brain's aptitude to pursue certain types of careers or subjects. The study and learning of subject matters, both uploads necessary information into short-term and long-term memory storage areas to optimize one's proficiency in pursuit of a career, but also enhances the network connections between processors, which further augments an individual's proficiency.

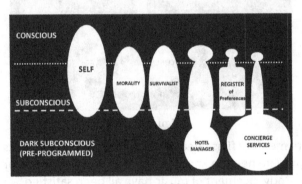

Figure 232

Summary of Human Brain Behaviors to include Self, Morality, Survivalist, Hotel Manager, Register of Preferences, Concierge Services.

A physician's brain generally utilizes the frontal lobes of the brain. Learning for physicians utilizes a STORE and RETRIEVE process. Physicians are generally excellent memorizers, training themselves to rapidly retrieve information that has been stored. During the process of memorizing vast quantities of information organized into elaborate decision trees are established in the brain of a physician. The more knowledge and the greater number of decision trees stored in a physician's brain, coupled with a rapid recall mechanism, generally translates into what is functionally required to be an effective clinical doctor. To be fair, every person is an individual and a physician must process medical problems in the context of differing parameters, which are unique to an individual including age, body mass index, and a wide spectrum of possible comorbidities; but the recent dictum to be followed have been what some would refer to as a cookbook approach to medical problems, where like problems are all treated similarly.

Engineers generally utilize the back portion of the brain nestled between the posterior parietal lobes the occipital lobes and the temporal lobes. The parietal lobes receive input from the peripheral sensors scattered throughout the skin covering the body. Inputs from the face, arms, hands, chest, back, legs, and feet flood into the parietal lobes. Right parietal lobe receives signals from the left side of the body, while the left parietal lobe receives inputs from the right side of the body. Occipital lobes receive input from the eyes via the occipital nerve that courses from the posterior of each eyeball to become embedded into the occipital lobes at the posterior of the brain. The temporal lobes, which hug each side of the human brain, contain processing centers for hearing, mathematics, and music. Engineers are generally 3D-spatial thinkers, working with concepts in three-dimensions because they utilize the portion of the brain between the lobes that receive input from the peripheral sensors of the body and eyesight, and combine this with the mathematic functions of the brain. Engineers learn mathematic algorithms and analysis skills in order to solve interesting and new problems as the world of technology evolves to higher states of complexity.

'Given sufficient education and the freedom of thought, the human mind is capable of solving any problem.'

Lane B. Scheiber II

POSTSCRIPT III

THE REASON
PRIME DIRECTIVE:
PROTECTING BLUEPRINTS
OF LIFE AT ALL COST

IN GENERAL, PHYSICS, LIKE other sciences is often studied as a matter of curiosity. Once developed, physics is often applied to the industrial sciences. Advancing physics has been critical in progressing the technologic presence of humanity in the world.

Physics is often shrouded in self-generated complexity. The exquisitely complicated mathematical equations tend to shield the average reader from understanding current principles of physics. Mathematical symbols, often unique to physics equations, help to keep the core principles of the study of physics hidden from the general populous.

Whether we wish to consider it or not, extinction of Mankind looms in our future. Numerous reasons why humanity would die off exist: A simple measurable wobble in the Earth's orbit causing the planet to be closer or farther away from the sun could result in a catastrophic increase or lowering of the planet's average surface temperature beyond what humans could survive in. Slow asphyxiation due to widespread use of the combustion engine. Irreversible contamination of the oceans' ecosystems due to the irresponsible discarding of plastic debris into the planet's waterways. Space debris, an asteroid only seven miles wide, struck the earth resulting in global extinction of the dinosaurs along with most of the surface life inhabiting the planet at the time. Given there is no autoregulation of growth for humans, humans run the risk of exhausting the planet's resources to the point of warmongering for resources. Many other fatalistic scenarios tantalize the imagination including (1) a catastrophic event will occur involving the surface of the planet such as a major movement of the tectonic plates, (2) a super volcano erupts, spewing enormous amounts of highly toxic ammonia and sulfur gases into the atmosphere, contaminating the breathable air, (3) a highly aggressive fatal pathogen will be generated either by nature or by humans, which will threaten the species, (4) we will encounter a more sophisticated alien race which wishes to eliminate the human race from the planet.

Less in the fantastic realm, more in the slowly perking of reality, is that either mankind or nature poisons the environment sufficiently that the earth is not able to sustain human life. Countries which exist in the northern latitudes of the earth are witnessing significant climate change as a progressive ongoing phenomenon, which is impacting their lives in real time. The trickledown effect of polluting the planet will eventually reach all of us.

Though the last scenario seems so farfetched it should only be considered in the chapters of a science fiction novel and certainly never be a concept entertained amongst the pages of a classical physics text; this opinion may itself be fatalistic.

Much of life on earth begins as a seed. It is not without the distinct possibility that the origin of life on earth is the result of the earth being seeded. Many known seeds are capable of existing for decades or longer before encountering the proper environmental conditions for a dormant seed to sprout life.

If Mankind were to suddenly be faced with the real possibility of extinction, there are not many viable choices for preserving life. Yet, as Mankind, we are the stewards of life. We are the guardians of the genetic blueprints of organic carbon-based organisms charged with making sure the essence of life does not become extinguished in the universe.

There is at least one effort, where seeds for a variety of plants have been stored in climate-controlled environment. Such an effort is dependent upon the surface of the planet not suffering some horrific catastrophe, which would damage the viability of the seed storage facility. The success of such a project also hinges upon the surface of the planet returning to a viable state to support life, before the timeline for the viability of the seeds expires.

Two realistic scenarios exist. One plan is the migration of the human species from planet earth by means of space flight. Earlier in this text, the concept of DORELIGHT galactic engines was presented and discussed. Even if such galactic engines could be constructed and implemented on a space craft, the question becomes, to what destination would humans travel to if suddenly the need to leave earth arose.

The closet stars is the trio of suns known as Alpha Centuri. The distance to the closest star in the Alpha Centuri system is 4.3 light years away. The fastest craft humans have constructed to date has already been discussed, being a satellite, which achieved a velocity of 36,000 miles an hour. So, if 36,000 miles an hour were the maximum speed of a spacecraft, it would take an ark 8,600 years to reach the nearest star. The speed of light is 720,000,000 miles an hour. If humans could build a spacecraft which could achieve the speed of light, it would still take four years and four months to reach the closest star to our Sun. As discussed earlier, a slip stream technology could become available to transport humans faster than the speed of light. How much faster than the speed of light will be determined as such technology evolves. If skilled engineers could master faster than the speed of light travel possibly utilizing slip stream technology, still at a speed four times the speed of light, it would take an entire year of deep space travel to reach the nearest boundaries of Alpha Centuri.

The nearest star, most likely would not provide humans a habitable environment to settle and flourish in by today's standards. By the limited information available, any of the planets associated with the three suns comprising the Alpha Centuri system are considered to be inhospitable to humans. To find a planet equal to the conditions found on earth would take a much longer voyage through space, most likely impossible to reach in a single human's lifetime.

So, if Mankind was suddenly threatened with extinction, the only viable option would be to save the blueprints of life written into the DNA code. One could store seeds or cellular structures deep in caves or man-made wells bored into the earth's crust. Due to increasing heat, as one penetrates deeper into the crust, the seeds of life or organic structures can only be hidden a limited distance under the surface of the Earth. For certain catastrophic failures, such as the collision of a sizeable celestial object with the Earth, concealing the seeds of life in the earth's crust may not guarantee survival of the blueprints of life.

To insure the blueprints of life would survive may require DNA be jettisoned from the Earth into deep space. Launching DNA in the direction of Venus, would not be a viable option due to the sulfuric acid content in the solar system's number two planet's atmosphere. An atmosphere containing any measurable amount of a powerful acid would circumvent any attempt life might have to survive. Transporting DNA to Mars would be equally fatalistic, given the fourth planet in the solar system is undesirably distant from the Sun, and Mars's molten core has gone cold. Without an active core, the fourth planet lacks a strong magnetic field to protect life from lethal radiation that rains down on the planet's surface from space. Mars's atmosphere is arid, predominantly carbon dioxide and often times clouded with dust storms which consume much of the planet. The remaining planets in our solar system have hydrogen and helium as surface gas atmospheres. Jupiter's moon, Europa, has at times been entertained as being able to support life in a geothermal water world suspected to exist under the moon's crust, still Jupiter is far more distant from the sun than Mars, and likely our Sun would exist only as a small dot in the moon's perpetually dark sky. There is no optimal habitat to cultivate and nourish any organic life as we know it, beyond the scope of the Earth's surface.

The reason why getting physics correct is high priority, is the critical need to understand the mechanics of magnetism and gravity, in order to at some point create seeds to preserve life. Such seeds would consist of organic genetic material housed inside a protective magnetic bottle. Suspended in a state of animation, such a seed could endure hundreds of billions of years of space travel. Eventually such seeds could transport the organic essence of life through the vastness of space to another planet in the galaxy, where the optimal environment for life exists, allowing life again to take root and flourish.

A complex understanding of physics is required to create the exterior of the seed, which would be utilized to jettison life into space and protect the organic DNA from the lethal radiation that would most certainly be encountered during space travel. Without physics, life will die out here on Earth, and that would be Mankind's greatest tragedy. The challenge before us, which may require the union of the greatest talents from across the spectrum of humanity, is to create seeds of life to insure life has a chance to survive beyond the scope of the Earth.

In essence, Mankind's circle of life may be similar to that of the life cycle of a dandelion. The seed of a dandelion plant falls to earth from the sky, sprouts jagged tooth shaped leaves and flourishes in some of the harshest conditions on the planet. The dandelion's characteristic yellow flower thrusts forward toward the sun and blooms. The dandelion plant is capable of reproducing sexually or asexually. The flower then converts into seeds constructed with specially crafted parachutes which carry the plant's generic code along the currents of the wind to distant places to repeat the cycle of life.

We may be destined to match the life cycle of the dandelion species, if the human race someday finds itself faced with extinction. The primary law of nature, transcending all other laws, is the protection of life's genetic code. If humanity is confronted with the possibility of a global extinction, we must utilize all of our talents and skills to create a seed containing life's genetic code, protect it with a magnetic shield and jettison such seeds out into the universe to hopefully locate and flourish on another planet similar to earth in a distant solar system. See Figure 233.

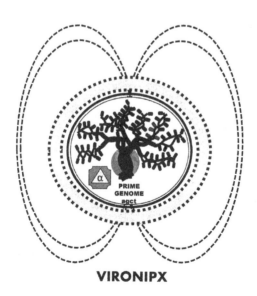

VIRONIPX

Figure 233

Vironipx Capsule

Humans have a mission, which is to extend the Prime Genome, primary programming instrument of life, beyond the boundaries of Earth. Humans possess the technical core memory files to accomplish this task. These files are systematically released to us as we progress toward the goal of launching the Prime Genome further into the galaxy. The skill of transforming inorganic molecules into organic life is the ultimate accomplishment any intelligent species could hope to participate in during the time line of their existence; it is their mark in the connectivity and continuum of the universe.

REFERENCES

1) Newton, I., The third book of opticks, Part I, *Opticks: Or, a treatise of the reflections, refractions, inflections & colours of light* (1704), (Cosimo Classics, New York, Republished 2007).

2) Aether, Brief history of the concept of aether, *Encyclopedianomadica* (Correa & Correa, 2005). Retrieved from http://encyclopedianomadica. org/English/aether.php.

3) Michelson, A. & Morley, E., On the Relative Motion of the Earth and the Luminiferous Ether, American Journal of Science, 34: 333–345, 1887.

4) Cecie, S., (2005). *Biology: Concepts and Applications.* Thomson Brooks/ Cole. ISBN 0-534-46226-X.

5) Waldman, G., *Introduction to light: the physics of light, vision, and color*(Dover ed.). Mineola: Dover Publications, 2002.

6) https://www.engineeringtoolbox.com/boiling-points-water-altitude-d_1344.html

7) https://en.wikibooks.org/wiki/High_School_Chemistry/Shapes_of _Atomic_Orbitals

8) https://education.jlab.org/qa/how-much-of-an-atom-is-empty-space. html

9) https://www.space.com/33527-how-fast-is-earth-moving.html

10) Rutherford, D., "Dissertatio Inauguralis de aere fixo, aut mephitico" (Inaugural dissertation on the air [called] fixed or mephitic), M.D. dissertation, University of Edinburgh, Scotland, 1772.

11) Lavoisier, A., *Elements of chemistry, in a new systematic order: containing all the modern discoveries.* Courier Dover Publications. p. 15., 1965, ISBN 0-486-64624-6.

12) Cavendish, H., Experiments on Air, Philosophical Transactions, 1785.

13) Thomson, J. J., "On the Structure of the Atom: an Investigation of the Stability and Periods of Oscillation of a number of Corpuscles arranged at equal intervals around the Circumference of a Circle; with Application of the Results to the Theory of Atomic Structure" (extract of paper). *Philosophical Magazine Series 6.* 7 (39): 237–265. 1904, doi:10.1080/14786440409463107.

14) Rutherford, E., Scattering of Alpha and Beta Particles by Matter and the Structure of the Atom, Philosophical Magazine, Series 6, Vol 21, May 1911, 669-688.

15) Bohr, N., "On the Constitution of Atoms and Molecules, Part II Systems Containing Only a Single Nucleus" (PDF). *Philosophical Magazine.* **26** (153): 476–502. 1913, doi:10.1080/14786441308634993

16) Electron Shell, en.m.wikipedia.org, accessed 20 January 2019.

17) Oerter, R., *The Theory of Almost Everything: The Standard Model, the Unsung Triumph of Modern Physics* (Kindle ed.). Penguin Group. p. 2. ISBN 0-13-236678-9, 2006.

18) Gell-Mann, M., A Schematic Model of Baryons and Mesons. *Physics Letters.* **8** (3): 214–215. Bibcode:1964, PhL.....8..214G. doi:10.1016/S0031-9163(64)92001-3

19) Zweig, G., "An SU(3) Model for Strong Interaction Symmetry and its Breaking", 1964, *CERN Report No.8182/TH.401.*

20) Zweig, G., "An SU(3) Model for Strong Interaction Symmetry and its Breaking: II". CERN Report No.8419/TH.412. 1964.

21) Fritzsch, H., Elementary Particles, Building Blocks of Matter, World Scientific Publishing Co, Pte, Ltd, Singapore, 2005.

22) Universe 101, Our Universe, WMAP Can Measure the Age of the Universe, NASA, wmap.gsfc.nasa.gov, accessed 20 January 2019.

23) http://environ.andrew.cmu.edu/m3/s2/02sun.shtml

24) Lemaître, G., (April 1927). "Un Univers homogène de masse constante et de rayon croissant rendant compte de la vitesse radiale des nébuleuses extra-galactiques"(PDF). *Annales de la Société Scientifique de Bruxelles* (in French). **47**: 49. Bibcode:1927ASSB...47...49L

25) *Overbye, Dennis (20 February 2017). "Cosmos Controversy: The Universe Is Expanding, but How Fast?". The New York Times. 2017*

26) Planck Collaboration (2015). "Planck 2015 results. XIII. Cosmological parameters (See PDF, page 32, Table 4, Age/Gyr, last column)". *Astronomy & Astrophysics.* **594**: A13.

27) Lawrence, C. R., (18 March 2015). "Planck 2015 Results" (PDF). Retrieved 24 November 2016.

28) "Age of the Earth". U.S. Geological Survey. 1997. Archived from the original on 23 December 2005. Retrieved 2006-01-10

29) Dalrymple, G.B., The age of the Earth in the twentieth century: a problem (mostly) solved, *Special Publications, Geological Society of London.* **190**(1): 205–221, 2001.

30) Manhesa, Gérard; Allègre, Claude J.; Dupréa, Bernard & Hamelin, Bruno, Lead isotope study of basic-ultrabasic layered complexes: Speculations

about the age of the earth and primitive mantle characteristics, *Earth and Planetary Science Letters.* **47** (3): 370–382. 1980.

31) Braterman, P., (2013). How science figured out the age of the earth, Scientific American, 2013, Archived from the original on 2016-04-12.

32) Jones, H. S., Huxley, T. H., *Proceedings of the Royal Institution of Great Britain*, Royal Institution of Great Britain, v. 38–39

33) Finley, D., Earth's Milky Way neighborhood gets more respect, National Radio Astronomy Observatory, 3 June 2013

34) Odenwald, S., Counting the Stars in the Milky Way, The Huffington Post. Archived from the original on August 1, 2014. Retrieved June 9,2014.

35) Martialay, M. L., The Corrugated Galaxy—Milky Way May Be Much Larger Than Previously Estimated, (Press release). Rensselaer Polytechnic Institute. Archived from the original on March 13, 2015.

36) Hall, S., (May 4, 2015). "Size of the Milky Way Upgraded, Solving Galaxy Puzzle", Space.com, Archived from the original on June 7, 2015. Retrieved June 9, 2015.

37) Ribas, I.; et al. First Determination of the Distance and Fundamental Properties of an Eclipsing Binary in the Andromeda Galaxy, Astrophysical Journal Letters. **635** (1): L37–L40. 2005.

38) CODATA Internationally recommended 2014 values of the Fundamental Physical Constants, Physical Measurement Laboratory of National Institute of Standards and Technology. Retrieved from http://physics.nist.gov/cuu/Constants/index.html.

39) https://solarsystem.nasa.gov/planets/earth/by-the-numbers

40) https://www.nasa.gov/audience/foreducators/k-4/features/F_Measuring_the_Distance_Student_Pages.html

41) https://undsci.berkeley.edu/lessons/pdfs/rutherford.pdf

42) Neary, G. J., The beta-ray spectrum of radium E, Proceedings of Royal Society of London, Series A, Mathematical and Physical Sciences, 175, 71-87 (1940).

43) Atomic Radii of Elements Data Page https://en.wikipedia.org/wiki/Atomic_radii_of_the_elements_(data_page)

44) http://www.chemicalelements.com/elements/n.html

45) Kendall, C., Periodic Table Nitrogen Isotopes, USGS, Isotope Tracers, https://wwwrcamnl.wr.usgs.gov/isoig/period/n_iig.html

46) Burns,K., Coupled Multi Group Neutron Photon Transport for the Simulation of High Resolution Gamma-Ray Spectroscopy Applications, Disseration, Georgia Institute of Technology, 2009, page 79.

47) Kinetic Diameter, https://en.wikipedia.org/wiki/Kinetic_diameter

48) Size of the Sun, Space.com, 31 October 2017.

49) The water in you, USGS, water.usgs.gov, accessed 20 January 2019.

50) How many atoms are in the human body? https://www.quora.com/How-many-atoms-are-there-in-the-human-body.

51) Planck, M., On the law of distribution of energy in the normal spectrum, Vol 4, p553-559, 1901.

52) Resolution 1, On the revision of the International System of Units, Resolutions of the 26th CGPM, Versailles, BIPM, bipm.org, 16 Nov 2018, accessed 23 Feb 2019.

53) Battye, R., Moss, A. Evidence for massive neutrinos from cosmic microwave background and lensing observations, Phys. Rev. Lett. 112, 051303, 6 February 2014. Retrieved from http://journals.aps.org/prl/abstract/10.1103/PhysRevLett.112.051303.

54) Henley, E., Garcia, A., The Weak Interaction, Chapter 11, Subatomic physics solutions manual, Third Edition, p30 (World Scientific Publishing Co, 2008).

55) NIST Reference, https://physics.nist.gov/cgi-bin/cuu/Value?gn

56) Brown, L., The idea of the neutrino, Physics Today, 31(9), p23-28, 1978.

57) Mertens, S., Direct neutrino mass experiments, Journal of Physics: Conference Series, 718 (2), arXiv:1605.01579, 2016.

58) Close, F., Neutrinos, Oxford University Press, 2010.

59) Jayawardhana, R., The neutrino hunters: The chase of the ghost particle and the secrets of the universe, Oneworld Publications, 2015.

60) Dodelson, S., Widrow, L., Sterile neutrino as dark matter, Physical Review Letters, 72(17), p17-20, 1994.

61) NASA Science, https://spaceplace.nasa.gov/dark-matter/en/

62) Periodic Table, https://en.wikipedia.org/wiki/Periodic_table

63) Griffiths, D., Introduction to Quantum Mechanics, Prentice Hall, Inc. Upper Saddle River, NJ, 1995.

64) Flowers, P., Theopold, K., Langley, R., 2.2 Atomic Orbitals and Quantum Numbers, chem.libretexts.org, last updated 15 Feb 2019.

65) Electron & Shell Configurations, Chemistry.patent-invent.com, accessed 2/16/2019.

66) Periodic Table: Copper, http://www.chemicalelements.com/elements/cu.html

67) Copper, https://en.wikipedia.org/wiki/Copper

68) Electron Configurations, https://chem.libretexts.org/Under_Construction/Essential_Chemistry_(Curriki)/Unit_1%3A_Atomic_and_Molecular_Structure/1.4%3A_Electron_Configuration_and_Orbital_Diagrams

69) Periodic Table: Gold, http://www.chemicalelements.com/elements/au.html

70) Jha, A., Why you can't travel at the speed of light, The Guardian, theguardian.com, 12 January 2014.

71) General Constants, link.springer.com, accessed 18 Feb 2019.

72) Scerri, E., The Periodic Table: Its Story and Its Significance; New Your City, NY; Oxford University Press, 2006.

73) Weeks, M., Discovery of the Elements; Easton, Pennsylvania, USA, Journal of Chemical Education, 6th ed, p 122, 1956.

74) Emsley, J., Nature's Building Blocks: An A-Z Guide to the Elements, New York, NY, Oxford University Press, 2011.

75) International Union of Pure and Applied Chemistry (IUPAC), Compendium of Chemical Terminology, 2 ed, Gold Book, 1997.

76) Soddy, F., The Origins of the Conceptions of Isotopes, Nobel Lecture, Nobelprize.org, 12 December 1922.

77) Barkla, C., The Spectral of Fluorescent Rontgen Radiations, The London, Edinburgh, and Dublin Philosophical Magazine and Journal of Science, 22 (129):396-412, 1911.

78) Moseley, H., The high-frequency spectra of elements, The London, Edinburgh, and Dublin Philosophical Magazine and Journal of Science, 26 (156):1024-1034, 1913.

79) Madelung, E., Mathematische Hilfsmittel des Physikers, Berlin, Springer, 1936.

80) Wieser, M., et al, Atomic Weights of the Elements 2011, IUPAC Technical Report, Pure Appl Chem, IUPAC, 85 (5), 1047-1078, published online 29 April 2013.

81) Cordero, B., Gomez, V., Platero-Prats, A., Reves, M., Echeverria, J., Cremades, E., Barragan, F., Alvarez, S., Covalent radii revisited, Dalton Trans, (21): 2832-2838, 2008.

82) Atomic Radii of the elements (data page), Atomic Radii, en.m.wikipedia. org., accessed 2 January 2019.

83) Atomic Radii, LibreTexts, chem.libretexts.org, 7 September 2017.

84) Bruno, T., CRC Handbook of Fundamental Spectroscopic Correlation Charts, CRC Press, 2005.

85) Electromagnetic Spectrum, https://en.wikipedia.org/wiki/Electromagne tic_spectrum

86) Electromagnetic Radiation, https://en.wikipedia.org/wiki/Electromagne tic_radiation

87) Dill, J., Lodestone and needle: the rise of the magnetic compass, http:// www.oceannavigator.com/January-February-2003/Lodestone-and-needle-the-rise-of-the-magnetic-compass/, Jan1, 2003.

88) Aurora, https://en.wikipedia.org/wiki/Aurora

89) Stuckeley, W., Memoirs of Sir Isaac Newton's Life, 1752, newtonproject. ox.ac.uk, published online September 2004.

90) Mahaffy, P.R., Webster, C. R., Atreya, S.K., et al, Abundance and isotopic composition of gases in the Martian atmosphere from the Curiosity Rover, Science, 341 (6143):263, 2013.

91) Zimmer, C., Earth's oxygen: a mystery easy to take for granted, The New York Times, 3 October 2013.

92) Hirt, C., Claessens, S.J., Kuhn, M., Featherstone, W.E., Kilometer-resolution gravity field of Mars: MGM2011, Planetary and Space Science, 67, (1): 147-154, 2012.

93) Weisenberger, D., How many atoms are in the world, Jefferson Lab, education.jlab.org. accessed December 15, 2018.

94) Mars Fact Sheet, nssdc.gsfc.nasa.gov, accessed December 15, 2018.

95) About Rainbows, The National Center for Atmospheric Research and UCAR Office of Programs, web.archive.org, accessed December 15, 2018.

96) Helmenstine, A., How to Calculate the Number of Atoms in a Water Droplet,Thoughtco.com, 8 January 2018.

97) Long, T., Nov. 8, 1895: Roentgen Stumbles Upon X-rays, wired.com, 8 Nov 2010.

98) Waters, H., The First X-ray, 1895, The Scientist, the-scientist.com, 1 July 2011.

99) Behling, R., Modern Diagnostic X-ray Sources, Technology, Manufacturing, Reliability, Boca Raton, FL, USA: Taylor and Francis, CRC Press, 2015.

100) Delchar, T., Physics in Medical Diagnosis, Springer, p 135, 1997.

101) Crooks Tube, The New International Encyclopedia, 5, p407, Dodd, Mead & Co., 1902.

102) Tyndall, J., The Electric Light, The Popular Science Monthly, Vol 14, March 1879.

103) Davis, L., Fleet Fire, Arcade Publishing, New York, 2003.

104) Henderickson III, K., The Encyclopedia of the Industrial Revolution in World History, Vol 3, p 564, Rowman & Littlefield, 2014.

105) Williams, H., Cassell's Chronology of World History, p 434-435, London, Weidenfeld & Nicolson, 2005.

106) Burns, E., Electric Lighting, The Story of Great Inventions, p 118-125, Harper & Brothers, 1910.

107) Keefe's, T., The Nature of Light, ccri.edu/physics/keefe/light.htm, 2007, last updated 27 Jan 2019.

108) Armaroli, N., Balzani V, Towards an Electricity-Powered World, Energy Environmental Science, 4, 3193-3222, 2011.

109) Brammer, G., van Dokkum, P., et al, A Remarkably luminous galaxy at z = 11.1 measured with the Hubble space telescope grism spectroscopy, The Astrophysical Journal, 819(2), 129, March 2016.

110) Cutnell, J., Johnson, J., Physics, New York, Wiley, p 468, 1997.

111) Zitzawitz, N., Davids, M., Physics: Principles and Problems, New York, Glencoe, p 308, 1995.

112) Siegfried, T., Einstein's Genius Changed Science's Perception of Gravity, Science News, Vol 188 (8), p 16, 17 Oct 2015.

113) Einstein, A., On a Heuristic Point of View about the Creation and Conversion of Light, Annalen der Physik, 17 (6), p132-148, 1905.

114) Nobel Prize Physics 1921, Nobel Foundation, nobelprize.org

115) O'Neil, I., How a Total Solar Eclipse Helped Prove Einstein Right About Relativity, Space.com, 29 May 2017.

116) Lights All Askew in the Heavens, Special Cable to the New York Times, London, Nov 9, New York Times, timesmachine.nytimes.com, 10 Nov 1919.

117) Don't Worry Over New Light Theory, New York Times, Archives, nytimes.com, 16 Nov 1919.

118) Einstein, A., Fundamental Ideas and Problems of the Theory of Relativity, Lecture Delivered to the Nordic Assembly of Naturalists at Gothenburg, nobelprize.org, 11 July 1923.

119) Wolchover, N., Can Anything Escape from a Black Hole?, Live Science, livescience.com, 15 Nov 2011.

120) Jones, H., Huxley, T, Proceedings of the Royal Institution of Great Britain, Royal Institution of Great Britain, v.38-39.

121) Balick, B., Brown, R, Intense Sub-Arcsecond Structure in the Galactic Center, Astrophysical Journal, 194(1), p265-270, 1 Dec 1974.

122) Xue-Bing, W., Wang, F., Fan, X.; et al., An Ultraluminous Quasar with a Twelve-Billion-Solar-Mass Black Hole at Redshift 6.30, Nature, 518 (7540), p512-5, 26 Feb 2015.

123) Einstein, A., Rosen, N., The Particle Problem in the General Theory of Relativity, Phys Rev, 48 (73), 1935.

124) Wormhole, en.m.wikipedia.org, accessed 27 Jan 2019.

125) Weyl, H., Feld und Materie, Annalen der Physik, 65 (14), p 541-563, 1921.

126) Weyl, H., Stanford Encyclopedia of Philosophy, plato.stanford.edu, accessed 28 Jan 2019.

127) Misner, C., Wheeler, J., Classical Physics as Geometry, Ann Phys, 2 (6), p 525, 1957.

128) Rothman, T., Was Einstein the First to Invent $E = mc^2$?, Scientific American, 333 (3), scientificamerican.com, 24 Aug 2015.

129) Bodanis, D., E=mc^2: A Biography of the World's Most Famous Equation, Bloomsbury Publishing, 2009.

130) Einstein, A., Does the Inertia of a Body Depend Upon Its Energy-Content?, Annalen der Physik, 18, p 639-643, 1905.

131) International Bureau of Weights and Measures, The International System of Units (330-331) (3rd ed), U.S. Department of Commerce, National Bureau of Standards, p 17, 1977.

132) The International System of Units, NIST Special Publication (330), U.S. Department of Commerce, August, 1991.

133) Joule, Unit of Energy Measurement, Encyclopedia Britannica, Britannica.com, accessed 29 Jan 2019.

134) Planck, M., Deutsche Physikalische Gesellschaft (German Physics Association), Max Planck 'Black Box Radiation' Lecture to DPG, 14 Dec 1900.

135) Planck's constant, Editors of Encyclopedia Britannica, Britannica.com, last updated 28 Dec 2018.

136) Marquardt, R., Meija, J., Mester, Z., et al. Definition of a mole, IUPAC Recommendation 2017, Pure Appl. Chem. Vol 90(1): 175-180, 2018. Degruyter.com.

137) Einstein, A., On the electrodynamics of moving bodies, Annals of Physics, 322(10), p 891-921, 1905.

138) Newton, I., Newton's Principia. Translated to English by Andrew Motte, Published by Daniel Adee, New York, 1846.

139) Havens, J., A brief history of the combustion engine, wiki. vintagemachinery.org, last modified 15 Jan 2015.

140) Naming Stars, International Astronomical Union, iau.org.

141) Lewin, S., What do we know about alpha centauri?, Space.com, 13 April 2016.

142) Tate, K., The nearest stars to earth, Space.com, 19 Dec 2012.

143) Yeager breaks the sound barrier, This day in history October 14, 1947, history.com, last updated 21 August 2018.

144) Warp drive, en.m.wikipedia.org, accessed 25 Feb 2019.

145) Carlson, C. The proton radius puzzle, Physics Department, College of William and Mary, Williamsburg, VA 23187, USA, February 19, 2015. Retrieved from https://arxiv.org/pdf/1502.05314.pdf.

146) Pohl, R., Antognini, A., Nez, F., et al. The size of the proton, Nature 466, 213-216 (2010).

147) Antognini, A., Nez, F., Schuhmann, K., et al. Proton structure from the measurement of 2S-2P transition frequencies of muonic hydrogen, Science 339, 417-420 (2013).

148) Nakamura, K., Hagiwara, K., Hikasa, K., et al. Review of particle physics, The Baryon Summary Table, Journal of Physics G: Nuclear and Particle Physics, Volume 37, article 075021. Retrieved from http://iopscience.iop.org/article/10.1088/0954-3899/37/7A/075021/pdf.

149) Yue, S. Y., On the radius of the neutron, proton, electron and the atomic nucleus, the General Science Journal. Retrieved from http://www.gsjournal.net/old/physics/yue.pdf

150) Henley, E., Garcia, A., The Weak Interaction, Chapter 11, *Subatomic physics solutions manual*, Third Edition, World Scientific Publishing Co, 2008, p30

151) Prout, W., Relation between the specific gravities of bodies in their gaseous state and the weights of their atoms, Thomson's Annals of Philosophy, 6, 321-330, 1815 (published anonymously)

152) Tretkoff, E., May, 1911: Rutherford and the discovery of the atomic nucleus, This Month in History, APS News, American Physical Society, Vol 15, No. 5, May 2006. Retrieved from http://www.aps.org/publications/apsnews/200605/history.cfm

153) Tretkoff, E., May, 1932: Chadwick reports discovery of the neutron, This Month in History, APS News, American Physical Society, Vol 16, No. 5, May 2007. Retrieved from http://www.aps.org/publications/apsnews/200705/physicshistory.cfm

154) Kopecky, S., Harvey, J., Hill, N...et al, Neutron charge radius determined from the energy dependence of the neutron transmission of liquid ^{208}Pb and ^{209}Bi, Physical Review C, Vol 56, No.4, 2229-2237, October 1997.

155) Auerbach, N., Neutron-proton radii in N≈Z nuclei. Retrieved from https://arxiv.org/ftp/arxiv/papers/1006/1006.2034.pdf.

156) Storti, R., Desiato, T., Derivation of fundamental particle radii: electron, proton, and neutron, April 10, 2015. Retrieved from http://documents.mx/documents/derivation-of-fundamental-particle-radii-electron-proton-neutron.html.

157) Wilson, D., The History of Pi. Retrieved from https://www.math.rutgers.edu/~cherlin/History/Papers2000/wilson.html.

158) Freud, S., Lecture XXXI, The Dissection of the Psychical Personality, New introductory lectures on psycho-analysis. The Standard Edition of the Complete Psychological Works of Sigmund Freud, Vol XXII (1932-1936): 1-182. Yorku.ca, last accessed 12 Mar 2019.

159) Netter, F., The Nervous System, The Netter Collection of Medical Illustrations,

160) Scheiber, A., The human computer, iuniverse, 2002.

161) Suh, H. A. (editor), XII.Inventions and experiments, Leonardo's Notebook, Black Dog & Laventhal Publishers, NY, 2005.

162) Leonardo da Vinci Inventions, da-vinci-inventions.com, last accessed 2 Mar 2019.

163) Mendeleev, D., The Periodic Table, Famous Scientists, famousscientists. org, last accessed 12 Jan 2019.

164) Mendeleev, D, en.m.wikipedia.org, last accessed 2 Mar 2019.

165) Pasteur, L., Lecture, University of Lille (7 Dec 1854), en.m.wikiquoate. org, last accessed 2 Mar 2019.

Printed in the United States
By Bookmasters